Beautiful Wreaths

40 HANDMADE CREATIONS THROUGHOUT THE YEAR

Melissa Skidmore

Skyhorse Publishing

Skyhorse Publishing books may be purchased in bulk at special discounts for sales promotion, corporate gifts, fund-raising, or educational purposes. Special editions can also be created to specifications. For details, contact the Special Sales Department, Skyhorse Publishing, 307 West 36th Street, 11th Floor, New York, NY 10018 or info@skyhorsepublishing.com.

Visit our website at www.skyhorsepublishing.com.

10 9 8 7 6 5 4 3 2 1

Library of Congress Cataloging-in-Publication Data is available on file.

Cover design by Qualcom
Cover photo credit by Melissa Skidmore

Print ISBN: 978-1-5107-4410-3
Ebook ISBN: 978-1-5107-4409-7

Printed in China

Dedication

I believe that each of us is created in the image of God.

"In the beginning, God created . . ." Genesis 1:1

We were created to create.

Thank you, Mom and Dad, for fostering that creativity in me; for showing me the love and warmth of home.

Thank you to my David Skidmore. This journey of life I am sharing with you is more than what I could have imagined. The way you support me, overlook the messes that my projects create, are so patient with me, and love me so completely is more than I deserve. I love you with all of my heart and am so thankful you asked me to be your wife all those years ago.

Thank you to my precious girls—Daisy, Anna Belle, and Lila Mae. You have given me one of the greatest joys in my life by making me a mom. Thank you for the ways you notice the little details of our home and for the way you have oohed and ahhed at each new wreath I've made. Y'all are so incredibly creative yourselves, and I *love* to watch y'all create!

Contents

ADORED
DECORATE
DESIGN*SPONGE at HOME

Introduction

I grew up in a creative home.

My parents wouldn't necessarily consider themselves creative, but my dad woke up every morning to a cup of coffee and a yellow legal pad where he would sketch ideas for things he wanted to build. His chosen material for creativity was wood. And the table saw was humming most weekends, as he worked to create his sketches that week.

My mom's canvas was our home. She was forever rearranging furniture, hanging pictures, making throw pillows, and sewing tablecloths. Our home was always a warm and welcoming place due to her loving attention to detail. I vividly recall the day she bought her first silk flowers. Our neighbor across the street walked over with an armful of flowers, announcing that these fine, high-quality silk flowers were on sale at our local department store. My mother and I packed ourselves into the Oldsmobile and took off. My mother had recently seen some beautiful grapevine wreaths, and she was ready to try one for herself. I remember the buckets loaded with flowers. As my mom handed me stem after stem, I could see she was getting more and more excited about the endless possibilities these realistic reproductions had to offer! She rushed home with her mind swirling with ideas. She sat down at the kitchen table, gathered her hot glue gun and wire snips, and began making her first wreath. When she was done, she proudly hung it on our front door.

The neighborhood ladies oohed and aahed over her creation—everyone wanted a wreath of their own. Before I knew it, my father had sketched out a design for a wreath-building stand and set up a little work area for her in the garage. I recall the day they set off for my uncle's farm, spending the day cutting grapevines and making their own wreaths. They came home with dozens of circular bound vines. My mom was giddy with excitement. She made wreaths for everyone we knew and made sure that every room in our house featured a different wreath. Our front door was never bare. It held new wreaths every season for as long as I can remember.

I am forever grateful to my parents for planting a creative spirit in me. And while I have no skills when it comes to working with wood, I take after my mom and treat my home like a canvas. More than twenty years ago, my husband, David, and I bought our first home. We were on a pretty tight budget, but I knew that I wanted my guests to feel welcome from the moment they arrived at our front door. And I knew that our home would not say "welcome" without a wreath! I scraped together ten dollars and headed to the local discount store. I came home to my kitchen table, got out my hot glue gun and wire snips, and got to work.

Our house has not been without a wreath ever since. We have also added three precious daughters to our home—Daisy, Anna Belle, and Lila Mae. I have strived to pass along the same creative spirit my parents had shared with me. As our girls got older and I had a bit more time to myself, I decided to go into the wreath-making business. I called my dad and had him build me a wreath making stand, and I set up my shop in my kitchen. I have now made and sold over three thousand wreaths.

The process of creating feeds my soul. I find such joy in it! Sometimes, the completed project is more than what I could have dreamed—and sometimes it is an epic fail! Most of the time, it falls somewhere in the middle. Let your joy come from the process of creating, and relish in your creation, rather than in the approval of others.

In these days of Pinterest, Instagram, and Facebook, there is such tremendous pressure for our homes, the parties we throw, and our general lives to look a certain way. There is so much comparison among our peers, and with that comparison comes anxiety and deep feelings of inadequacy. My mom never had that pressure. The meals she cooked, the beautiful projects she completed for our home, and the parties she hosted were just for the people she loved. That is how I try to live—living fully with my tribe. Living fully in the moment and not caring if the pictures will be good for social media. Yes, I love creating beautiful things, and I love for my home to be warm and welcoming. But, my main motivation beyond followers and likes is to create a place where my people enjoy being and where new people will want to come. I desire this for my girls—I want the homes they create to be warm and welcoming for the sake of their loved ones,

and not for the sake of keeping up an image for others. And I want this for you, too!

My intention with this book is to encourage creativity, just as it was encouraged in me. The wreaths you create will not look exactly like the ones in this book—and that's perfectly fine! You *should* make any changes you like to make the wreath your own. My hope is that you can sit with this book and be inspired to create something beautiful for your own home or as a gift for a friend.

Happy wreath making!

Basic Supplies and Skills

Wreath-Making Supplies

Artificial Flowers and Greenery

I choose to use artificial flowers and greenery for almost all of my wreaths. I love that I can make a wreath once—and then use it for years! Artificial flowers and greenery are gaining popularity, and they are available in a wide variety of stores. My first stop is always local craft stores, which tend to have the largest variety of flowers and greenery at the best prices. You can also find them in home decor stores, big box discount stores, and dollar stores. I am very selective with the materials I use and always make sure that they can pass for the real deal. Since I like to examine and touch the flowers to make sure they look real, I do most of my shopping in person. When I can't find what I'm looking for in the store, I shop online. My favorite stores are afloral.com, save-on-crafts.com, Michaels.com, and HobbyLobby.com.

Wire snips

This is a tool you will use with virtually every wreath you create. Most artificial floral stems have a sturdy wire down the center. You will want a quality pair of wire snips so you can easily cut that wire. You will also use wire on occasion to secure garland and heavier objects to your wreath.

Needle-nose pliers

These will come in handy when attaching the wire hanger to the wreath and when working with fine wire to attach objects to the wreath.

Glue gun

This is another tool that you will use with almost every wreath. I prefer a high-temperature glue gun as it seems to have better holding power than low-temperature glue. I also like to keep a thick piece of cardboard or paper plate under the glue gun to catch drips.

Glue sticks

Make sure to buy glue sticks that are for your type of glue gun. They come in short sticks, long sticks, and coils. I prefer a long, thick, high-temp glue stick.

Hand-held pruners

Most grapevine wreaths need a bit of trimming before you can work with them. You may be tempted to cut the stray vines with your wire snips, but do not do this as you will ruin your wire snips!

Wire

There are three types of wire I use—20-gauge 18-inch stem wire for making wire hangers on the back of each wreath, 18-gauge 18-inch stem wire for making bows, and 26-gauge green floral paddle wire when attaching garlands to wreaths.

Scissors

It is good to have two pairs of scissors—one for removing tags from anything and everything and a higher quality pair for cutting ribbon. Dedicate the higher quality scissors to ribbon only, or they will get dull and you will not be able to cut a pretty edge on your ribbon.

Wreath forms

There are several types of wreath forms. The most popular are grapevine, straw, wire, and Styrofoam. They come in many different shapes and sizes, and just about any wreath can be adapted to a different size. I would recommend measuring your space before you choose the size of your wreath form. Remember that the finished product will typically be 1 to 4 inches wider than the wreath form you start with.

All of these basic supplies can be found at craft stores or hardware stores.

Tips for Using Grapevine Wreaths

Grapevine wreaths are one of my favorite supplies to use, but you need to know a thing or two about them. I would not recommend ordering these online. Due to the great variance in each wreath, I recommend buying them in person so you can select the right ones.

- Many grapevine wreaths are warped. You want a grapevine wreath that will lie flat when you lay it on a flat surface.
- Your grapevine wreath should have a tight weave of vines. If the gaps are too large, there is nothing for the hot glue to adhere to when you are attaching flowers and greenery.
- You want each of the vines to be of a fairly consistent smaller diameter. There will be some variance, but you don't want really large stems woven through as this will make the wreath heavy.
- Any grapevine wreath will have a lot of random twigs. I always trim these extra twigs off to restore the true circular shape of the wreath. I also try to remove as many of the dried leaves as I can. This is a messy job that I would recommend taking outside!

Tips for Making Wreaths

- It is best to make your wreath in an upright position (not lying down on a flat surface). The wreath should be at about eye level to allow you to correctly position the flowers and other elements. An easy way to do this is to buy an over-the-door hanger, hang the wreath on a door or a kitchen cabinet, and work there. You will want to hang your wreath near a work surface for your supplies and an outlet for your glue gun.
- Some of the basic wreath designs in this book can be easily dressed up for a holiday by simply changing or adding a bow. They can also be turned into a seasonal wreath by attaching a small bunny, pumpkin, or Christmas ornament with some wire. These items can be removed after the holiday and your wreath can continued to be used all year round.
- Remember that any wreath that is left outdoors will begin to weather. Keep that in mind when you are choosing elements for each wreath. Some artificial flowers are actually made for the outdoors and will be marked accordingly. I have had good luck using regular artificial flowers for outdoors, as long as the wreath gets some protection from the weather.
- As you make each wreath, remember that you are in charge. Keep bending and fluffing and working with the leaves and petals to get them where you want them to be! Sometimes the smallest adjustment to a leaf or the placement of the flower petals make a big difference in the final product.

How to Hang a Wreath

There are several ways to hang a wreath. I like to choose the least visible method so that the beautiful wreath is the star of the show.

- Over-the-door hangers are good to hang wreaths on most standard doors. This style comes in clear plastic, a variety of metals, and highly decorative designs.
- Magnetic hooks are perfect for metal doors.
- Suction-cup hooks are a good choice for glass doors. Make sure that the suction cup is large enough to support the weight of your wreath.
- Removable adhesive hooks work for a variety of surfaces.
- A small nail is perfect if you are hanging your wreath on drywall or a wood surface.

- If you want to hang your wreath on your vinyl siding, there are hooks made specifically for it.
- As for those of you who have a storm door, I would like to ask a favor: would you *please* hang the wreath on the outside of the storm door and not sandwiched between the door and storm door? Yes, the wreath will receive some weather, but you will get to enjoy the wreath fully until it is time to be replaced.

How to Make a Wire Hanger

This is a basic technique for creating a wire hanger and attaching it to your wreath. I have found that I can adapt this simple concept to create a hanger for every wreath, even nontraditional ones.

Supplies

20-gauge floral stem wire

Needle-nose pliers

ONE

Lightly bend the floral stem wire in the middle.

TWO

Twist the ends of the wire together to create a loop about 5" long.

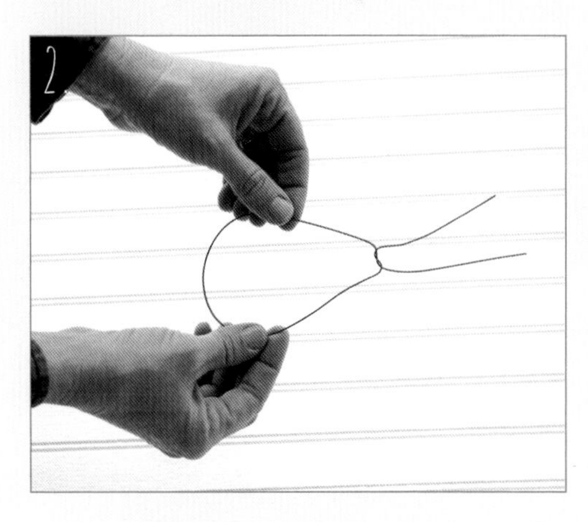

THREE

Twist the two ends of the wire down the loop on each side.

FOUR

Attach the hanger to the wreath by sliding the loop through a gap in the vines in a secure position at the top of the wreath. This will vary depending on the type of wreath form you are using. Lead the loop with the twisted edge.

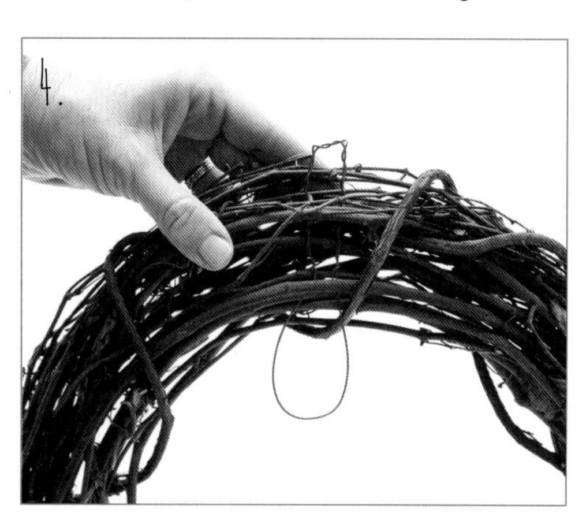

FIVE

Bring the clean end of the loop up and through the twisted edge of the loop.

SIX

Pull the clean edge of the loop snug and twist at the base to secure it.

SEVEN

You now have a loop to hang your wreath from.

How to Make a Classic Bow

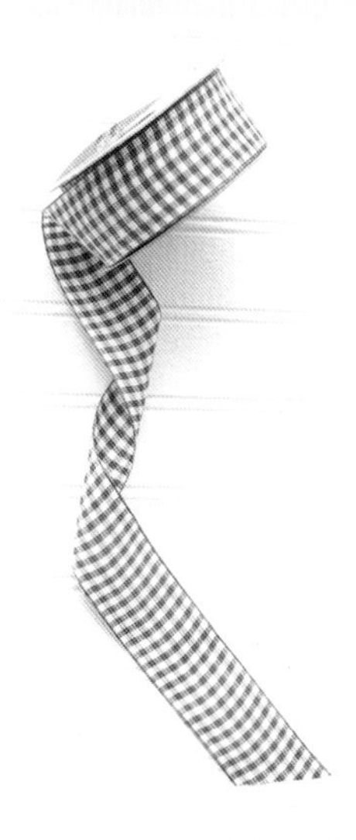

My mom taught me how to do this years ago. I held my fingers out for her countless times so she could tie perfect bows, and now my girls do the same for me.

Supplies

1–2 yards of your favorite ribbon

Scissors

A friend

ONE

Have your friend hold their fingers as shown. The width they hold their fingers apart is how wide the bow will be. Choose a person who can hold their hands steady! Place the ribbon over their fingers as shown.

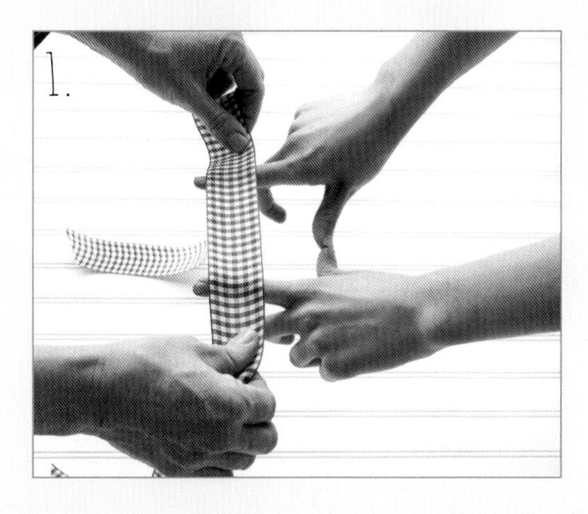

TWO

Cross the ends under one another.

THREE

Pull one of the ends toward your friend and the other end back to yourself.

FOUR

Fold the piece that you pulled toward your friend back over the top and down to meet the other end.

FIVE

Tie a knot with the two ends. Make sure that both of the loops are even on either side of the knot before you tie it down.

SIX

Tighten the knot.

SEVEN

Fluff the bow and it is ready to use!

How to Make a Pom-Pom Bow

This is my go-to bow. I make it more than any other! It takes a bit of practice, but once you perfect this bow, you can use it on everything.

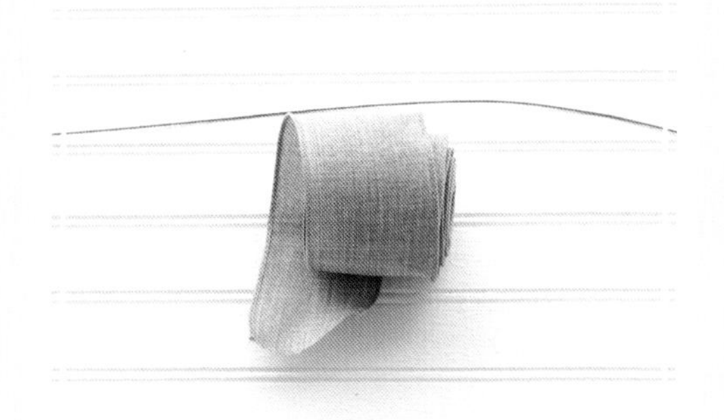

Supplies

At least 4 yards of ribbon

18-gauge floral stem wire

Needle-nose pliers

Scissors

ONE

Decide how long you want the tail of the bow to be and make your first twist in the ribbon.

TWO

Measure out about 8" more of ribbon and fold the first loop back to the center. Twist the ribbon and stack on the other twist. Hold the center tight between your thumb and fingers.

THREE

Alternate sides and continue this process until you have 3 loops on each side.

FOUR

Make the second tail the same length as the first and cut the ribbon at an angle.

FIVE

Using the needle-nose pliers, twist the floral wire stem around the center. You will use these wire stems to attach the bow to the wreath, so leave the wires long.

SIX

Fluff the loops of the bow and arrange them as shown.

Spring

"My favorite weather is bird-chirping weather."
—Terri Guillemets

Spring is my absolute favorite time of year. I want to fling all the windows open and let the sun shine in! I feel life being breathed back into me when I see those first daffodils blooming. The days start getting a little bit longer and those temperatures begin to rise, sending my spirit soaring. The hope of things to come energizes me. I feel a spring in my step after the restful season of winter. This collection of spring wreaths will help you to celebrate this season with its fresh beginnings—full of baby showers, weddings, garden parties, and Easter celebrations. Spring is the season of *life*!

Note: The supplies pictures in the following projects are just a representation of what the supplies look like. Not all of them show the exact number of items needed to make the wreath.

Spring Floral Wreath

I love a recipe that needs only a few ingredients and very little time to make—and then everyone wants seconds and thinks you're a hero for cooking it up. That is exactly what this wreath is. Very few supplies, very little time spent, with plenty of wow factor!

Supplies

3 bushes white ranunculus

Wire hanger (p. 5)

22" grapevine wreath

Tools

Wire snips

Needle-nose pliers

Hot glue gun

Hot glue sticks

Instructions

ONE

Cut each ranunculus bush into individual stems. Each stem should be about 3" long. Slide the leaves to the top of the stem close to the bloom.

1.

TWO

Attach the wire hanger to the top of the grapevine wreath (p. 5). Hang up the wreath and continue working. Apply hot glue to the end of the stem. Starting just left of the center of the top of the wreath, insert the first three blooms in a cluster.

2.

THREE

Continue to apply hot glue to the end of the stems and cluster more of the ranunculus tightly together.

3.

FOUR

Add more flowers to the top-left corner to make it the fullest part of the wreath. Begin to fade the flowers toward the ends. Tuck leaves behind blooms where needed. I like to see as many of the blooms as possible.

FIVE

Find the perfect home for this wreath and wait for the oohs and aahs to follow!

Grapevine Bunny

I'll never forget the day my dad brought home my pet bunny. I named her Princess, and he built a bunny house for her. Unfortunately, Princess was quite the escape artist, and my dad spent many an afternoon chasing her all over our neighborhood. My mom and I would stand at the kitchen window and watch and laugh. We had no idea that my dad was that nimble and could run so fast! Bunnies make me smile to this day.

Supplies

14" round grapevine wreath

16" oval grapevine wreath

24-gauge brown floral wire stem

Brown grapevine wire

Green faux moss bendable vine wire

Spanish moss

Wire hanger (p. 5)

Adhesive-backed moss

Wired ribbon pom-pom bow (p. 10)

Greenery and berry stems

Tools

Wire snips

Hot glue gun

Hot glue

Needle-nose pliers

Instructions

ONE

Attach the round grapevine wreath to the oval wreath using the 24-gauge brown floral wire stem.

TWO

Cut the brown grapevine wire into two 16" pieces. Shape each piece into a bunny ear.

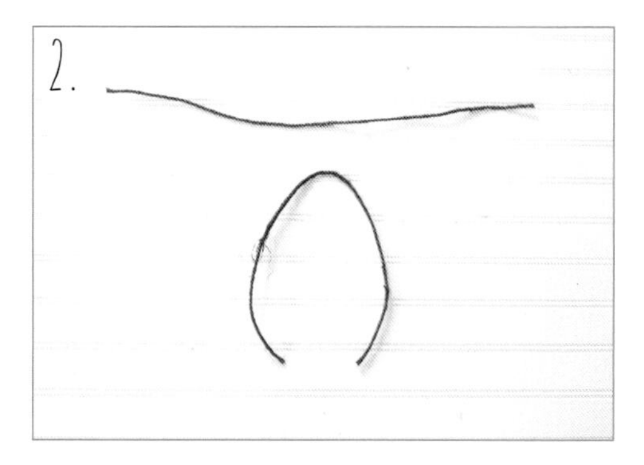

THREE

Insert the ends of each ear into the top of the grapevine form about 1" deep and wrap the ends of the wire around a few of the grapevine pieces.

FOUR

Cut the green faux moss bendable vine wire into two 20" pieces and wrap around each brown wire bunny ear.

FIVE

Shape the Spanish moss into a fluffy tail. Apply a lot of hot glue to the moss and hold it onto the grapevine until the glue is dry. Be careful not to burn your fingers while doing this! At this point, attach the wire hanger to the top of the wreath (p. 5) and move the bunny to an upright workspace.

SIX

Cut a strip of the adhesive-backed moss and remove the adhesive back. Place it over the seam of the two wreaths to cover the joint.

SEVEN

Using the wire stem on the pom-pom bow, secure it at the neck of the bunny, off-centered.

EIGHT

Apply hot glue to the end of the greenery and berry stems and insert them behind the bow.

NINE

Your bunny is ready to welcome your family and friends all spring long.

Watering Can Arrangement

After a long dreary winter, nothing thrills my soul like the first blooms of tulips. These little beauties let us know that the garden is waking up and that warmer weather is on its way. This arrangement holds all of the promise of spring!

Supplies

2 white tulip bushes

Oval-shaped vintage watering can

3–4 pieces floral foam

Tools

Wire snips

Serrated knife

Instructions

ONE

Cut each tulip stem from the bush, leaving the stems as long as you can. Often, you should be able to slide the leaves up and down the stems. You can adjust these leaves as you arrange them in the watering can.

TWO

Fill the watering can with the floral foam. Fill it to about 1" from the top of the watering can. Use a serrated knife to cut the foam if needed to fit into the can. The blocks I used happened to fit this watering can perfectly!

THREE

Hang up the watering can from its handle. It might tilt at a slight angle, which adds to the organic feel of this wreath.

FOUR

Insert the tulips into the foam in the watering can as shown.

FIVE

Continue adding tulips.

SIX

Pack the tulips in tight. Slightly bend the stems to make them look more natural.

SEVEN

Find the perfect place to hang this arrangement so that it easily welcomes spring!

Boxwood Wreath

I've always heard that everyone should own a little black dress. I've never owned one myself, because my life never seems to require it, but a great pair of jeans is an essential in my closet. They are so easy to wear, and they go with everything. This wreath is the equivalent of those jeans. It will go in any space and can be used all year round. Hang it on a door or a mirror or the back of a chair. The possibilities are endless!

Supplies

2 boxwood bushes

1 greenery garland bush

Wired burlap ribbon pom-pom bow (p. 10)

14" round grapevine wreath

Wire hanger (p. 5)

Tools

Wire snips

Hot glue gun

Hot glue sticks

Needle-nose pliers

Instructions

ONE

Clip the boxwood bush into separate 3" pieces as shown.

1A.

1B.

2.

TWO

Apply hot glue to the top of each stem, creating a barrier so that the boxwood does not slide off the top of the stem.

THREE

Clip the greenery garland bush into separate 3" pieces.

FOUR

Attach the pom-pom bow to the top of the wreath by inserting the wire into the grapevine wreath and twisting it at the back to secure the bow. If you wait and add the bow after the boxwood, your bow will not lay correctly.

FIVE

Apply hot glue to the ends of the garland greenery pieces and insert them into the grapevine form until the form is covered.

Attach the wire hanger (p. 5) and move the wreath to an upright position to continue.

SIX

Apply hot glue to the ends of the boxwood pieces and insert 5 of them around the bow.

SEVEN

Continue placing the boxwood pieces around the center of the wreath.

7A.

7B.

EIGHT

When the center of the wreath is full, begin inserting the stems into the sides of the wreath to fill it out.

Flip the wreath over and use the wire snips to clip any stems that protrude from the wreath.

8.

NINE

This wreath is ready to go! Get creative with where you are going to hang it!

Monogram Wreath

I am a Southern girl, born and raised in the Tennessee. And we Southern girls love a monogram. Our mommas monogram everything—our silver cups, our towels, our sheets, our hairbows . . . the list goes on and on! So, naturally, I had to make a monogram wreath to give a nod to the Southern mommas of the world.

Supplies

Large papier-mâché letter

1" ribbon

Floral foam

26-gauge floral paddle wire

Adhesive-backed moss

2 ranunculus bushes

Tools

Serrated knife

Wire snips

Instructions

ONE

Using the serrated knife, cut the top off the front of the papier-mâché letter.

TWO

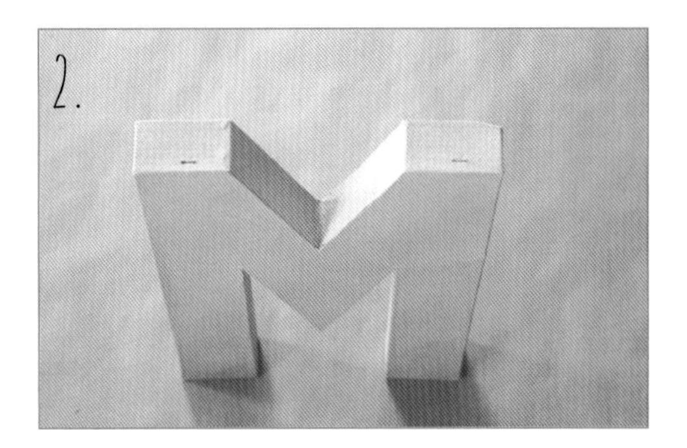

Using the serrated knife, cut two small slits in the top edge of the letter.

THREE

Cut a piece of 1" ribbon about 24" long and insert each end into the slits at the top of the letter. Knot the ends of the ribbon to secure them.

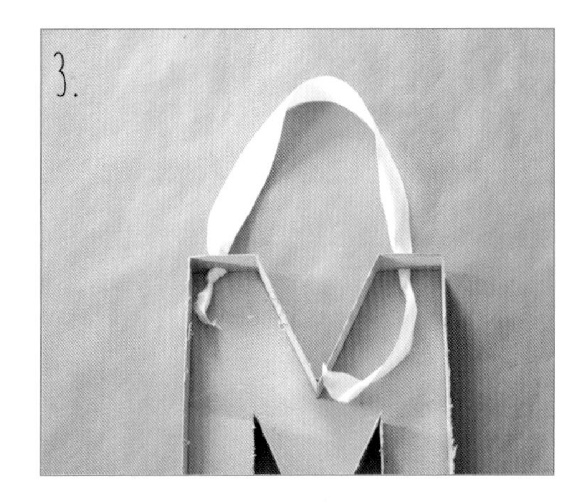

FOUR

Using the serrated knife, cut the floral foam to fit inside the hollows of the letter and fill it up. You want the foam to sit flush with the edge of the papier-mâché.

FIVE

Wrap the 26-gauge floral paddle wire around different parts of the entire letter to secure all of the pieces of foam. Then secure the wire by wrapping and twisting it back around itself.

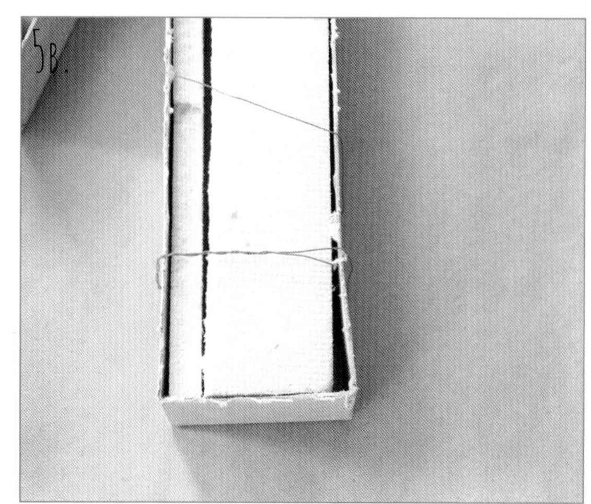

SIX

Cut the adhesive-backed moss to fit the sides of the letter and stick it on.

6A.

6B.

SEVEN

Clip the ranunculus bushes into pieces. If there are leaves, clip those as well. Leave about ½" stems on each piece.

7

EIGHT

Begin at the bottom and insert each ranunculus stem into the foam.

NINE

Continue working your way around the letter. Insert the leaves around the edge and concentrate the blooms toward the center of the letter. This will allow the letter to keep its shape better.

TEN

This wreath is ready to hang! Simply loop the ribbon over your choice of wreath hanger. It's so cute on a child's bedroom door or to welcome friends to a baby shower!

Rose Heart Wreath

I love that a simple heart shape speaks love. Make this wreath for the special people you want to spread a little love to; you don't have to save it for that one special holiday. Hang it in your family room, on a nursery door, or on your front door. Everyone who sees it will know what you're saying.

Supplies

Trailing flower bush

Berry bush

Wire hanger (p. 5)

Heart-shaped grapevine wreath

Tools

Wire snips

Needle-nose pliers

Hot glue gun

Hot glue

Instructions

ONE

Clip the trailing flower bush and the berry bush so the stems are 4–6" long.

TWO

Attach the wire hanger (p. 5) to the heart-shaped wreath and hang it up. Continue to work from there.

THREE

Apply hot glue to the ends of the berry stems and insert them into the grapevine form.

FOUR

Bend the stems as needed to make them take the shape of the heart.

FIVE

Apply hot glue to the ends of the trailing flower stems and insert them into the grapevine form.

SIX

Place a flower where the curved sides of the heart shape meet in the middle. This will define the shape better. You want a natural, carefree look, but you still want to see the shape of a heart!

SEVEN

Place another flower at the bottom of the heart where the straight ends meet. This one can trail off a bit, but will still add that definition.

EIGHT

Now hang it in a place where you can speak love to those you care for! This wreath hangs on a gallery wall in my bonus room.

Magnolia Wreath

I have always loved magnolias. Where I grew up, these trees would get enormous. The thick leathery leaves could be used as fans to cool you off in the summer—they also made the perfect "plate" for mud pies. I have many fond memories of these Southern beauties.

Supplies

2 magnolia leaf garlands

18" round grapevine wreath

26-gauge floral paddle wire

Wire hanger (p. 5)

Tools

Wire snips

Needle-nose pliers

Hot glue gun

Hot glue

Tip: This wreath can be made with any garland of your choice! Just apply the basic principles of this wreath-making process. If you choose a garland with smaller leaves, you might need three or four garlands to make the wreath nice and full.

Instructions

ONE

Often, the garland will have one main branch of greenery and several smaller branches stemming off from the main branch. Clip the smaller branches from the main branch.

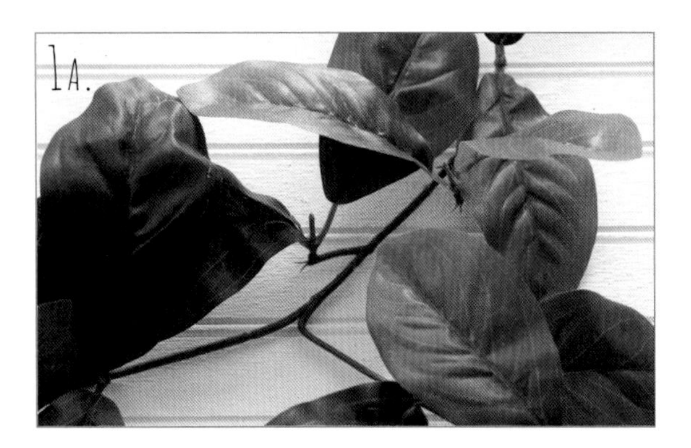
1A.

1B.

TWO

Starting at the bottom-left of the grapevine wreath, secure the 26-gauge floral paddle wire to a few of the grapevine pieces. Insert the end of the magnolia leaf garland into the grapevine wreath and wrap the wire around the garland and the grapevine wreath.

2.

THREE

Begin wrapping the floral paddle wire around the grapevine wreath and the magnolia leaf garland to secure the garland to the wreath. Be careful not to wire any of the leaves down, only the main branch.

FOUR

Continue to wrap until the entire length of magnolia leaf garland is secured to the grapevine wreath. The garland will probably overlap a bit.

FIVE

Turn the wreath over and secure the end of the floral paddle wire around a piece of the grapevine with a knot.

SIX

Clip the second magnolia leaf garland into pieces. Make each cut right above a magnolia leaf, leaving about 1" of stem on each piece.

SEVEN

Attach the wire hanger to the top of the grapevine wreath (p. 5). Hang the wreath in an upright position to finish it. The wreath will look very uneven as of now.

EIGHT

Apply hot glue to the end of the magnolia leaf garland stems and insert them into the grapevine form. Fill in the gaps until the wreath is full.

NINE

Now this beauty is ready to find its new home!

Egg and Moss Wreath

There is something so charming about the idea of having chickens in my backyard and sending my girls out to gather their fresh eggs every morning. But it will have to remain just a dream in my mind. My Anna Belle is scared to death of birds; we think it has something to do with the time when Anna Belle was six and we were in San Diego—a bird flew down and snatched her cheeseburger from her little hand. Until she recovers from her fear, I will just have to imagine how cute this little wreath would be on my chicken coop door.

Supplies

Preserved Spanish moss

14" round straw wreath

26-gauge floral paddle wire

Wire hanger (p. 5)

Lightweight eggs, for crafting, in 3 or 4 colors

Tools

Wire snips

Needle-nose pliers

Hot glue gun

Hot glue

Instructions

ONE

Loosely form the preserved Spanish moss to the straw wreath.

Secure the 26-gauge floral paddle wire to a couple of strands of the straw and begin wrapping the wire around the Spanish moss and straw wreath form to secure them together. Continue wrapping the Spanish moss around the wreath until it is fully covered.

Secure the wire by looping it under a few pieces of straw in the back and tying a knot in the wire.

1A.

1B.

1C.

TWO

Attach the wire hanger (p. 5) to the back of the wreath. Hang up the wreath in an upright position and continue.

Choose one of the colored eggs and hot glue it to the Spanish moss. Continuing gluing on eggs of the same color, spacing them out evenly.

THREE

Choose the next color and hot glue the eggs in place, spacing them out evenly.

Continue with the next color, and then the next. I love how eggs come in so many beautiful, soft colors!

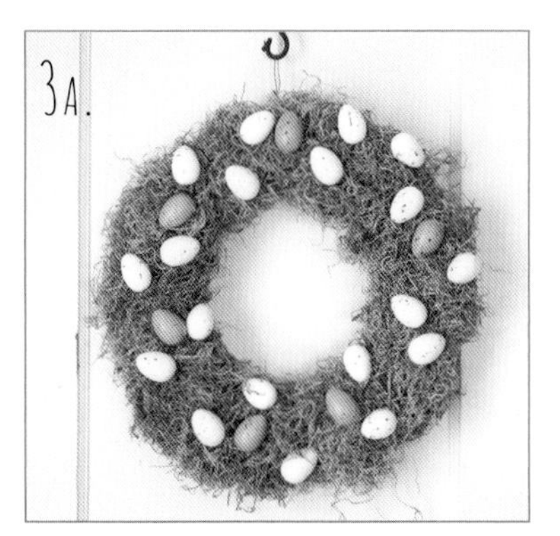

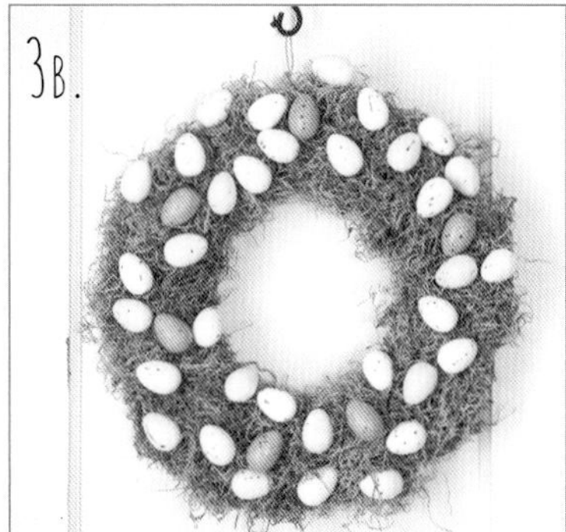

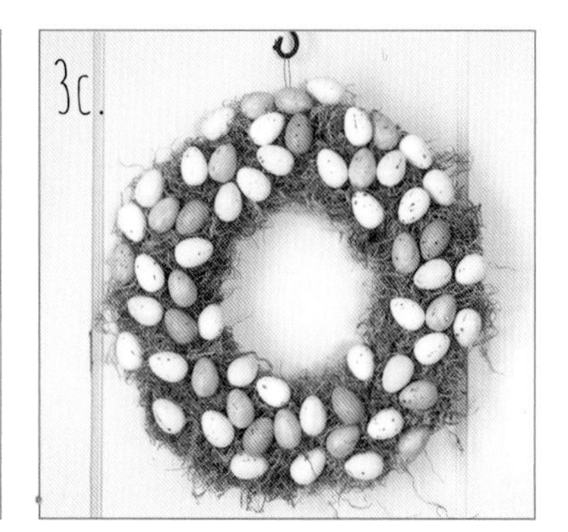

FOUR

Hang up this wreath in your chicken coop—or any other place you like to welcome in spring!

Summer

"I love how summer just wraps its arms around you like a warm blanket."

—Kellie Elmore

Summer . . . this momma loves summer. Schedules seem to disappear. The days get long. Children can stay up late at night catching lightning bugs and watching movies. Days are spent swimming and playing and trying to find cool places to hang out when the days get too hot! A few yard projects might get done. Weeks are filled with camps and sleepovers; families visit to grill out. Ice cream is a daily treat. I am that strange mom who wishes that school would never start. I love soaking up these days with my precious family.

Summer Floral Swag

I was delighted when I found out that there is a beautiful white hydrangea that shares my middle daughter's sweet name—Anna Belle. The year she was born, my mother-in-law gave me an Annabelle hydrangea to plant. This plant was neglected badly for the first several years of its life because I was too busy raising little girls to tend to flowers. Yet, somehow, it has flourished and produces an abundance of luscious white blooms year after year! There is nothing prettier than a hydrangea arrangement.

Supplies

1 bush trailing greenery

1 bush small white ranunculus

3 stems large white ranunculus, with a few flowers on each stem

2 stems small pale green hydrangea, with several flowers per stem

3 stems large white hydrangea

Long twig swag base with a large loop at the top

Wired ribbon cream pom-pom bow (p. 10)

Tools

Wire snips

Hot glue gun

Hot glue

Instructions

ONE

Clip the trailing greenery bush, small white ranunculus bush, large white ranunculus stems, and small pale green hydrangea stems. Keep each piece 3–4" long. Clip the 3 large white hydrangea stems to about 1". Clip the hydrangea leaves from the stem.

TWO

Hang up the swag. Apply hot glue to each piece of the trailing greenery bush and insert them into the swag until it is fully covered.

THREE

Start at the bottom. Apply hot glue to the end of the small pale green hydrangea stems and insert them into the swag.

FOUR

Hot glue the large white ranunculus stems above the small pale green hydrangea. Cluster some of the larger blooms together and scatter the smaller ones.

FIVE

Apply hot glue to the end of the large white hydrangea blooms and cluster them together at the top of the swag.

SIX

Hot glue the hydrangea leaves around the large white hydrangea blooms.

SEVEN

Hot glue the small white ranunculus bush stems around to fill in gaps.

EIGHT

Add the wired cream pom-pom bow to the top by inserting the bow's wires into the swag and twisting them together at the back.

NINE

Welcome friends and family to your home with this swag full of white flowery goodness.

Herb Wreath

My oldest, Daisy, is getting interested in gardening. We planted the most beautiful herb garden this year, with big plans of cooking with the fresh herbs and drying them. But instead, we just enjoyed looking at our plants, smelling them, and keeping them clipped and watered. Maybe next year we will actually dry them and use them for this wreath! Until then, here are some amazing artificial herbs that would be perfect for a herb wreath.

Supplies

2 each of 6 small bushes of artificial herbs, with varying leaf sizes and colors

Wire hanger (p. 5)

14" grapevine wreath

Tools

Wire snips

Hot glue gun

Hot glue

Needle-nose pliers

Instructions

ONE

Each bush will be different in how you need to clip them down to a workable size. If the leaves are adjustable, push all of them to the top of each stem and clip the stems so you have about a 1" stem.

TWO

Some of the bushes will need to be clipped apart piece by piece. Figure out the best way to get small pieces from each bush with about 1" stems that can be hot glued into the grapevine wreath.

THREE

Attach the wire hanger to the back of the grapevine wreath (p. 5) and hang up the wreath. Continue to work. Begin with one type of herb. Hot glue the end of each stem and place it into the grapevine in a cluster. Make sure to cover the outer and inner edge of the wreath.

FOUR

Pick another herb that has a different color and leaf texture. Hot glue the end of the stems and insert them into the grapevine. Cluster this together close to the first cluster of herbs.

FIVE

Continue with another type of herb, making sure again to vary the leaf color and texture. You see the pattern here now. Continue until you are done hot gluing all the herbs.

SIX

Hang it in your kitchen and enjoy! Maybe this will even inspire you to try your own herb garden this summer.

Fern Wreath

We visit a camp every summer in the mountains of New York, where the woods are loaded with ferns of every variety. They all mesh together to make the most beautiful covering on the ground. I imagine if I could gather those ferns and bring them home, I would make a wreath like this. New York, this wreath is for you!

Supplies

3 fern bushes, with varying leaf styles and color

Preserved moss

14" grapevine wreath

Wire hanger (p. 5)

Tools

Wire snips

Hot glue gun

Hot glue

Needle-nose pliers

Instructions

ONE

Separate each of the ferns into single fronds.

TWO

Hot glue the preserved moss to the upper right side of the grapevine wreath and the bottom left side of the grapevine wreath.

THREE

Add the wire hanger to the back of the grapevine wreath (p. 5) and hang up the wreath. Continue working. Start with the fern that you have the most of—in my case, it was the maidenhair fern. Apply hot glue to the end of the fern stems and insert them into the top and bottom sections covered with preserved moss. Add them to the outer edge of the preserved moss as well. Keep the look natural by allowing the fern fronds to bend in different directions.

FOUR

Begin working with the second type of fern and cluster it closely together at the bottom left of the grapevine wreath.

Mirror the top right section of the grapevine wreath with the second fern. The top of the grapevine wreath should be a reflection of the bottom of the grapevine wreath.

FIVE

Add the third type of fern to the bottom left and top right of the wreath. This will create layers of ferns similar to the natural ferns growing in the woods.

SIX

This wreath can be hung in many places. Place it on walls or doors where you need great texture and good memories of the woods.

Moss Wreath

When my husband, David, and I got married, I was so excited about decorating our first house! I loved the idea of polished cotton florals, while David loved burlap plaid. We met in the middle with neutral and I began to fall in love with linen and moss; I had plenty of moss DIY projects in that first house! Today, I still love a beautiful moss project.

Supplies

3 bags preserved moss

14" straw wreath

Floral pins

5 assorted types of mosses, in varying colors and textures*

Wire hanger (p. 5)

Tools

Hot glue gun

Hot glue

Needle-nose pliers

Scissors

**Tip*: Various assortments of mosses are often sold together in single bags.

Instructions

ONE

Separate the preserved moss into clumps that are about 2–4" in diameter. Apply hot glue to the back of each clump of preserved moss and hold it around the straw wreath until the hot glue is dry. Warning: the hot glue will seep through the moss, so work carefully and do not use more hot glue than necessary—you do not want it seeping through the moss and burning your hands!

TWO

Keep gluing the preserved moss to the front and sides of the straw wreath until it is covered. It is okay if there are some thin places here and there; the other mosses will cover it. The back of the wreath does not need to be covered.

THREE

Using floral pins, take small pieces of the assorted mosses and begin placing them around the wreath. Stick the floral pin through the moss and into the straw wreath. Reserve several pieces of the thicker moss to be used later in step 5 to cover the floral pins that show through the moss.

FOUR

Space the assorted mosses evenly around the wreath.

FIVE

Use a thick moss to cover up the floral pins that still show. Hot glue the thick moss over the floral pins.

SIX

Attach the wire hanger to the back of the wreath (p. 5) and hang, so you can see what needs to be trimmed and where more moss needs to be added.

SEVEN

Grab your scissors and give the stray moss pieces a little trim.

EIGHT

Wherever there are bare spots, add assorted moss here and there using either floral pins or hot glue until all of the spots are covered.

NINE

Hang this beauty in a place in your house that needs some lovely texture and warmth. You can also display it on a coffee table.

Hanging Basket Arrangement

Our summers are crazy busy. We leave for camp the first week the girls get out of school and are in and out of town all summer long! But whenever we are home, I love for everything to be beautiful and welcoming. (It makes the mountains of laundry easier to bear.) This arrangement can be thrown together in no time at all, making it a go-to during the summer months!

Supplies

7 matching long flower stems

Flat-backed basket*

Floral foam

Preserved moss

Wire hanger (p. 5)

1 yard wired ribbon (the same ribbon the bow is made from)

18-gauge floral stem wire

Wired ribbon classic bow (p. 7)

Tools

Wire snips

Needle-nose pliers

Scissors

Serrated knife

**Tip*: There are so many different styles of these flat-backed baskets available in craft shops and decorative home stores. They come in all shapes, sizes, and finishes. Just make sure the back is totally flat so it hangs properly.

Instructions

ONE

Measure the length of the long-stemmed flowers alongside the basket. Trim each long flower stem to within a couple of inches of the full length of the basket.

TWO

Fill the basket with floral foam. Use a serrated knife to cut the foam to fit inside the basket. Fill it to within a couple of inches from the top.

THREE

Cover the floral foam with the preserved moss. Attach the wire hanger to the back of the basket (p. 5) and loop the matching ribbon through the wire hanger. Hang the basket with the ribbon loop from your wreath hanger of choice.

FOUR

Insert the first 2 long flower stems in the middle of the foam.

FIVE

Insert the next two long flower stems beside the first two, spaced out as shown and slightly shorter than the first two.

SIX

Insert the next two long flower stems beside the first four. Space out and adjust the stems so they are slightly shorter than the others.

SEVEN

Using the 18-gauge floral stem wire, feed it through the knot on the bow and attach it to the basket by twisting the wire to the wicker.

EIGHT

How simple was that? And the flowers can easily be changed out for any season or occasion!

Garden Hose Wreath

Summers are hot here in Tennessee. When the girls were little, it was always a treat to pull out the garden hose and let them play in the water! They'd wash their bikes, fill pails and watering cans, and stay busy as little bees. Garden hoses represent sweet memories of laughter and playing for me.

Supplies

6' garden hose

26-gauge floral paddle wire

1 wispy floral stem

Pom-pom bow (p. 10)

Wire hanger (p. 5)

Tools

Wire snips

Needle-nose pliers

Instructions

ONE

Coil the garden hose into the size that you want it to be as a wreath. Arrange the hose in such a way that it will lay flat. Wrap the 26-gauge floral paddle wire around the top of the hose several times. Do not clip the wire at this point.

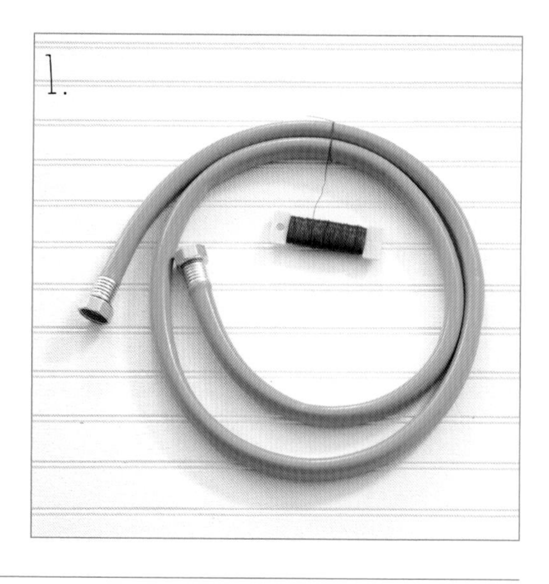

TWO

Clip the wispy floral stem into 3 sections. Wrap the floral paddle wire around each wispy flower stem several times, securing it to the top of the hose. When arranging the wispy flower stems, leave a gap for the pom-pom bow to go on afterward; the bow will lay flatter if there aren't flowers under it.

THREE

Wire the pom-pom bow onto the top of the hose. Make sure the bow is large enough to cover the stems and the wires connecting it all together. Attach the wire hanger to the back of the wreath (p. 5).

FOUR

Find a place to hang this super summery wreath!

Lavender Wreath

Sometimes life is crazy and calendars are full—and you crave a dose of simplicity. Lavender exemplifies simplicity. This wreath requires very little skill and time. It is so simple, yet beautiful. Create and hang this in your house in no time at all!

Supplies

3 bushes artificial lavender

18" grapevine wreath

Wire hanger (p. 5)

Tools

Wire snips

Needle-nose pliers

Hot glue gun

Hot glue

Instructions

ONE

Clip each stem from the artificial lavender bush. Leave a 1" stem to hot glue and insert into the grapevine form.

TWO

Attach the wire hanger to the top of the wreath (p. 5) and hang. Continue working. Hot glue the end of the artificial lavender stem and insert into the grapevine form. Use 3 stems of lavender per row. Start at the bottom left of the wreath.

THREE

Add three more stems of lavender about 1" away from the first row of stems.

Continue to glue more rows of lavender stems, following the shape of the grapevine wreath, until you complete the circle.

If you see some gaps that need more lavender stems, fill them in with more lavender.

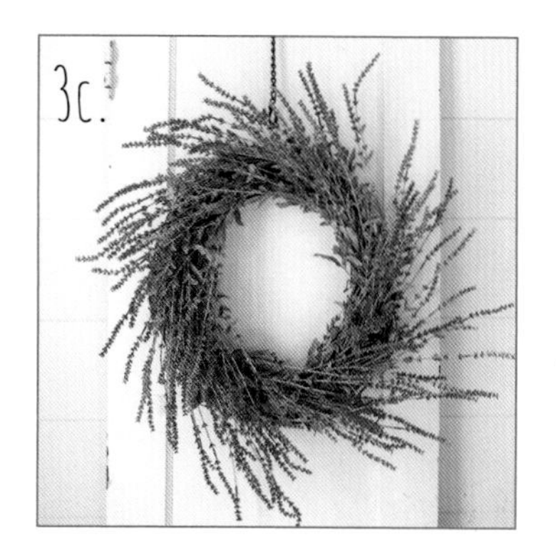

FOUR

This wreath would be perfect hanging over your tub in the bathroom or anywhere you need a bit of simplicity in your life!

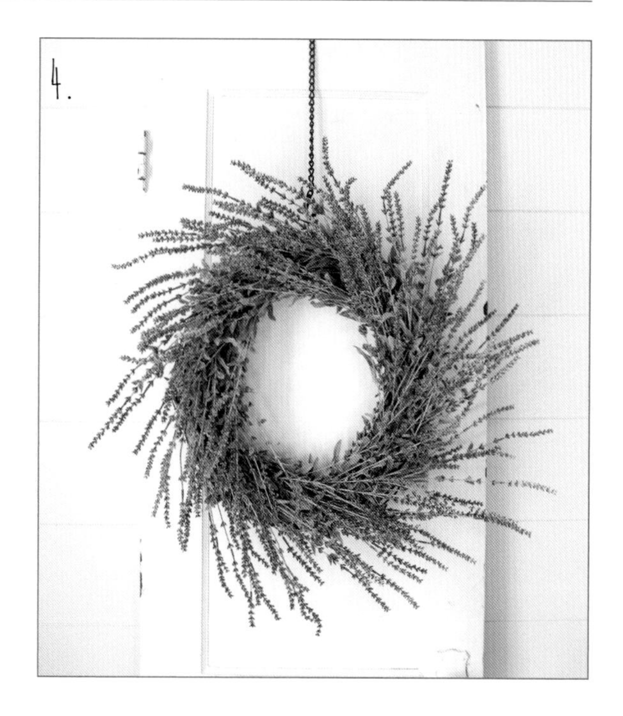

Clay Pot and Succulent Wreath

When I proudly showed my loving, supporting husband this new wreath, his exact words were, "Wow. That's so creative. I never could have thought of that." I was left to wonder at his words! This new creation is the perfect addition to any garden shed.

Supplies

24-gauge brown floral wire stem

11 3" tiny clay flower pots

24" wispy twig wreath

Floral foam

Wire hanger (p. 5)

11 faux succulents

Tools

Wire snips

Needle-nose pliers

Serrated knife

Instructions

ONE

Insert the 24-gauge brown floral wire stem through the hole in the clay pot from the bottom and out the top.

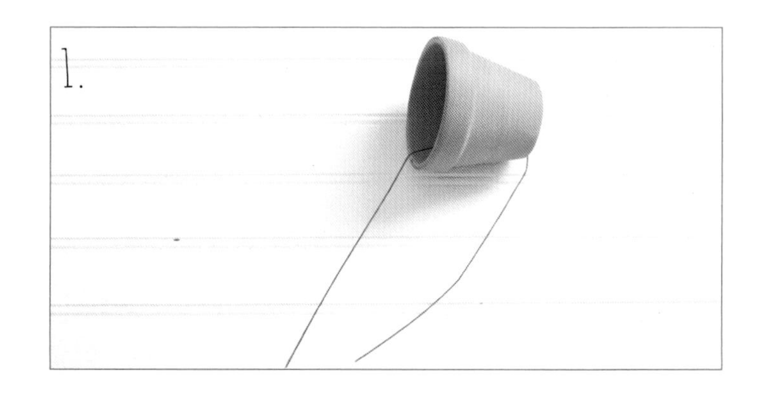

TWO

Secure the floral wire stem with the clay pot to the inner circle of the wispy twig wreath by wrapping the wire around the twig wreath's sturdy branches.

THREE

Continue adding the clay pots to the wispy twig wreath. Make sure that the pots are turned in different directions, but that none of them are upside down. Adjust the clay pots as needed to get them to fit securely.

FOUR

Add clay pots until the inner circle of the wreath is full. Using a serrated knife, cut the floral foam to fit each clay pot.

FIVE

Attach the wire hanger to the wispy twig wreath (p. 5). Hang it up and continue to work. Fill each pot with the floral foam.

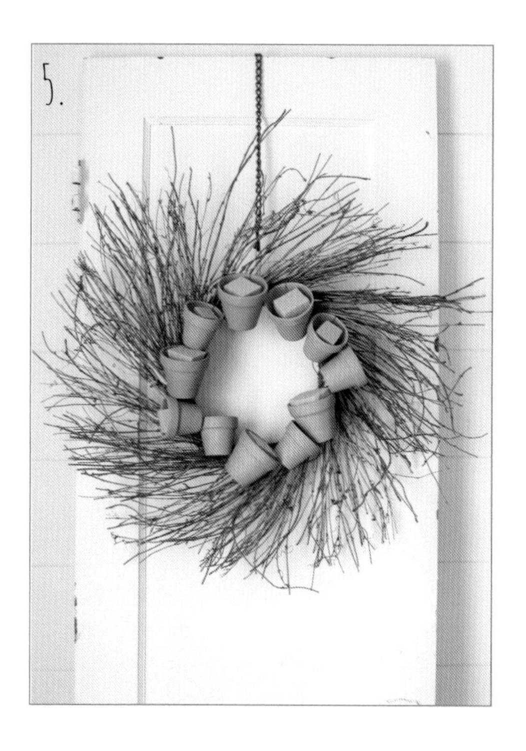

SIX

Using the wire snips, trim each succulent stem to fit into the pots.

SEVEN

Insert each succulent into a pot. Space the succulents in such a way that the colors and textures are balanced.

EIGHT

This wreath is right at home in a garden shed or on the gate of a back fence. Enjoy!

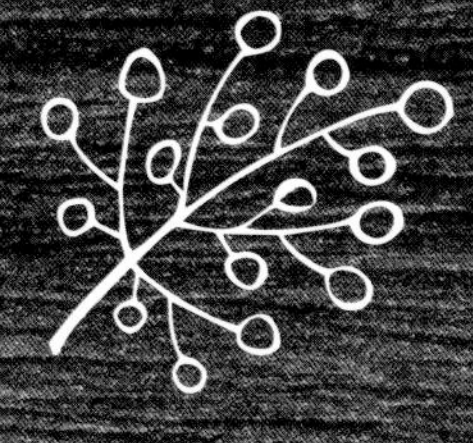

Fall

"Autumn is the mellower season, and what we lose in flowers we more than gain in fruits." —Samuel Butler

Autumn is always a welcome season after the heat and chaos of summer. The temperatures begin to fall. There are leaves to rake, pumpkin patches to visit, and big family dinners to plan. After a summer of constant movement, it is refreshing to think about resting at home and hosting the people we love. During early fall, our family takes our annual beach trip, where I have plenty of time to sit on the shore and dream up all the fall gatherings I plan to have. This fall collection of wreaths represents this season of home.

Wheat Wreath

While this wreath is not hard to make, it will require a bit of patience. Those little wheat stalks aren't as strong as you'd hope they would be, so be prepared to break a few of them. As you work, think happy thoughts about the golden wheat stalks that represent the beauty of one of the most glorious times of the year! Just keep clipping and gluing. The results will be worth it!

Supplies

3 bundles dried wheat

Wire hanger (p. 5)

14" straw wreath

Tools

Wire snips

Needle-nose pliers

Hot glue gun

Hot glue

Instructions

ONE

Clip each stem of dried wheat, leaving about a 2" stem.

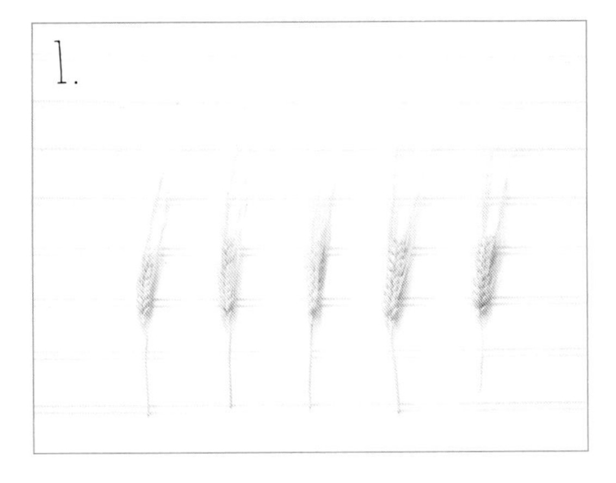

TWO

Attach the wire hanger to the top of the wreath (p. 5). Hang up the wreath and continue to work.

Apply hot glue to the ends of the wheat stems and slide them under the clear wrap that holds the straw wreath together. (This clear fishing line should already be on the straw form when you buy it.) If you push the straw down with one finger, you can slide the wheat under the fishing line more easily. Work gently.

THREE

Add the wheat stems to the wreath one row at a time. Each row should hold about 4 stems. Continue around the wreath, remembering to be patient. The little stems might break, but the wreath will be worth it in the end! Follow the curve of the straw wreath form and keep the angle of the wheat stems in such a way that the shape stays round.

FOUR

Go back and add more stems of wheat in bare spots. You can use some of the broken pieces for this.

FIVE

Hang this wreath to remind you of the golden goodness of fall.

Tennessee Spinner Gourd Wreath

I'm a Southern girl, specifically a Tennessee girl. Born and raised in West Tennessee I moved to East Tennessee when I got married, and now I've been settled for the last twenty years in Middle Tennessee. Tennessee has really been a most blessed place to have grown up in—and now to raise my own girls. These little Tennessee spinner gourds are just so cute and are my shout-out of love to my home state!

Supplies

Wire hanger (p. 5)

14" grapevine wreath

26-gauge floral paddle wire

Preserved moss

95 Tennessee spinner gourds

Tools

Wire snips

Needle-nose pliers

Hot glue gun

Hot glue

Instructions

ONE

Attach the wire hanger to the top of the grapevine wreath (p. 5). Secure the end of the 26-gauge floral paddle wire around a piece of the grapevine wreath.

TWO

Place the preserved moss on top of the grapevine wreath and wrap the floral paddle wire around the preserved moss and grapevine wreath to secure them together.

THREE

Continue adding preserved moss and wrapping the floral paddle wire around the wreath until it is covered.

Twist the end of the floral paddle wire to a piece of grapevine to secure in place.

FOUR

Hot glue the Tennessee spinner gourds onto the preserved moss.

Cluster the gourds together tightly. Continue to hot glue them on until the wreath is completely covered.

FIVE

Hang up your wreath and enjoy!

Fall Floral and Pumpkin Wreath

I am a fan of neutral colors. My girls make fun of me, saying that my favorite color is actually no color at all. The first time I saw a white pumpkin, I was charmed—be still my neutral-loving heart . . . I knew I needed to make a wreath with them immediately.

Supplies

1 frosted eucalyptus garland

18" grapevine wreath

26-gauge floral paddle wire

1 peach rose bush, with 5 large roses

1 white ranunculus bush, with 8–10 flower clusters

1 stem of cream berries, with 6–8 clusters of berries

1 beige heather bush, with 12–14 stems

1 eucalyptus bush

3 white 3" pumpkins on picks

Wire hanger (p. 5)

Tools

Wire snips

Needle-nose pliers

Hot glue gun

Hot glue

Instructions

ONE

Attach the frosted eucalyptus garland to the grapevine wreath by inserting the eucalyptus garland a few inches to the left of the top-center of the grapevine wreath. Twist the 26-gauge floral paddle wire around one of the grapevine twigs and the start of the eucalyptus garland.

TWO

When you wrap the floral paddle wire around the eucalyptus garland and the grapevine wreath, make sure to only catch the base of the eucalyptus garland with the wire and do not catch the leaves.

THREE

Continue wrapping the floral paddle wire around the eucalyptus garland and grapevine wreath in a clockwise direction until the garland is about to turn the circle and start up the side of the wreath. You don't want the leaves of the eucalyptus garland to be turned upside down, so clip the garland at the turn. Continue wrapping the wire around the grapevine wreath until you reach the starting point.

FOUR

Insert the tip of the clipped eucalyptus garland at the starting point again. In a counterclockwise direction, wrap the wire around the garland and grapevine wreath back to where the garland ended in step 3. Clip off the excess garland. Tie a knot in the floral paddle wire to secure it.

FIVE

Clip the excess frosted eucalyptus garland into individual pieces. These will be used later in step 9 to fill in the gaps of the wreath.

SIX

Clip the peach rose bush, white ranunculus bush, stem of cream berries, beige heather bush, and eucalyptus bush so the clippings have approximately 2" stems. Do not clip the wire stems in the white pumpkin picks. You will use these wire stems to attach the pumpkins to the wreath.

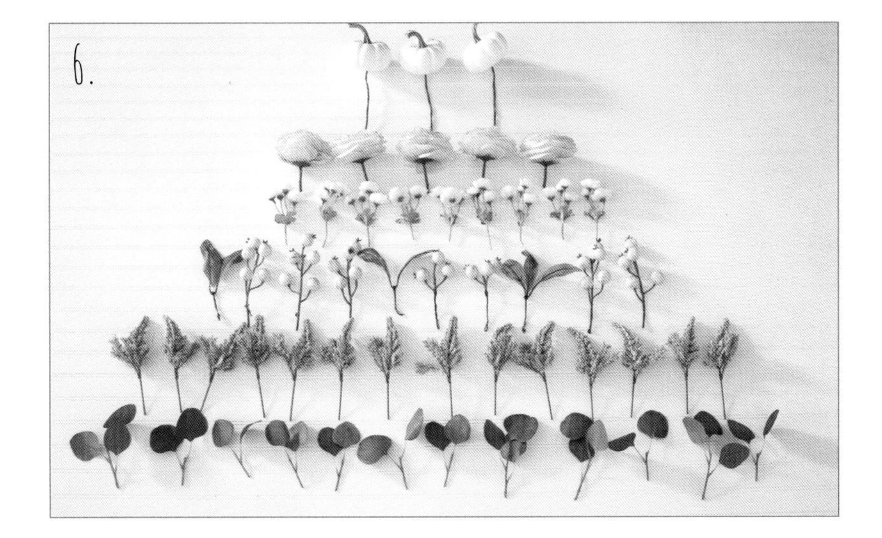

SEVEN

Attach the wire hanger to the top of the wreath (p. 5) and hang up the wreath. Continue working. Apply hot glue to the stems of the peach roses and insert them into the grapevine form. Cluster three of them together in the bottom-left corner of the wreath and the other two in the top-right corner.

EIGHT

Attach the white pumpkin picks to the grapevine wreath by inserting the wire stems into the wreath and twisting the wire around a piece of the grapevine. Place the first pumpkin in the top-left corner, opposite the two peach roses. Place the second pumpkin about 3 inches down from the first pumpkin. Place the last pumpkin evenly between the cluster of 2 peach roses and 3 peach roses.

NINE

Apply hot glue to the small, clipped stems from the eucalyptus garland and insert them into the grapevine wreath to fill in the empty spaces.

TEN

Apply hot glue to the stems of the white ranunculus bush and insert into the grapevine wreath. Place them evenly around the wreath.

ELEVEN

Apply hot glue to the cream berry stems and insert into the grapevine wreath. Place them evenly around the wreath. If your cream berry stem had leaves, add the leaves as well to fill in empty spaces.

TWELVE

Apply hot glue to the beige heather stems and insert into the grapevine. Place some of the beige heather stems on the outer edge of the wreath and others on the inner edge. This will add some movement to the wreath.

THIRTEEN

Apply hot glue to the eucalyptus stems from the eucalyptus bush and insert into the grapevine. These stems will be used to fill in any spaces and to round out the wreath.

FOURTEEN

Hang this wreath on the front door to welcome your guests all season long. It will make them fall in love with the neutrals of fall!

Gold Hoop Wreath

I have more of a traditional farmhouse style when it comes to wreaths, but my girls are pushing me to love modern designs. When they showed me these adorable gold hoop wreaths, I knew I had to give one a try! And of course, I added my own traditional flair to it.

Supplies

Eucalyptus bush

Wispy greenery stem

5" peony bloom

26-gauge floral paddle wire

10" gold hoop

Twine

Tools

Wire snips

Needle-nose pliers

Hot glue gun

Hot glue

Instructions

ONE

Trim off 4 stems from the eucalyptus bush into 6" pieces. Trim the wispy greenery stem into several pieces. Remove the peony bloom from the stem and discard the stem.

1.

TWO

Attach the end of the 26-gauge floral paddle wire to the 10" gold hoop by wrapping it around the hoop several times. Add the first stem of 6" eucalyptus by twisting the wire around the stem and gold hoop. Then, attach another piece of eucalyptus to the other side of the gold hoop, facing the opposite way, by twisting more wire around the stem and gold hoop.

2A.

2B.

THREE

Add the wispy greenery on top of the eucalyptus by twisting the wire around the stems and gold hoop.

FOUR

Add hot glue to the back of the peony bloom and hold it in place at the base of the gold hoop. The peony bloom will cover the wire.

FIVE

Loop the twine through the hoop and tie a knot at the top of the twine. Hang the wreath from the twine. Bend the stems as needed to follow the circular shape of the gold hoop.

SIX

Find the perfect home for this simple wreath!

Cornucopia Wreath

Oh, the bounty of fall! I load the girls up in the Suburban and we head on over to our local nursery, filling our carts with pumpkins of all colors and sizes, gourds, mums, and ornamental cabbages. I love the textures and rich colors of fall. Let your imagination run with this wreath; the fall possibilities are endless.

Supplies

4" x 8" piece of chicken wire

6–8" cornucopia*

18" grapevine wreath

18-gauge floral wire stems

Wire hanger (p. 5)

2 greenery and twig bushes

2 large pine cone and berry stems

2 gourd picks

6 sedum picks

Tools

Wire snips

Needle-nose pliers

Hot glue gun

Hot glue

*Tip: these come in a variety of materials; choose the one you like best.

Instructions

ONE

Fold the chicken wire and fit it inside the cornucopia.

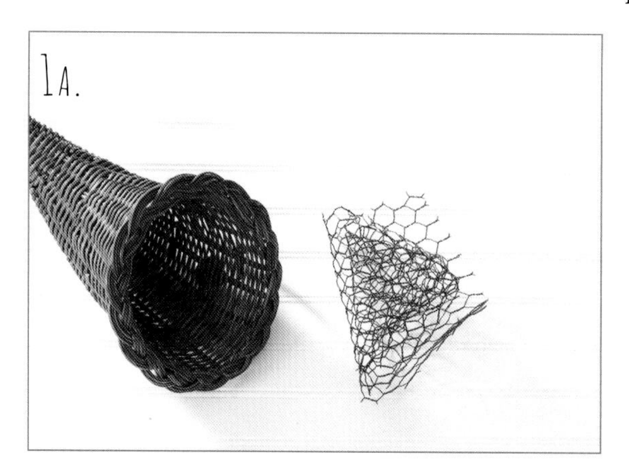

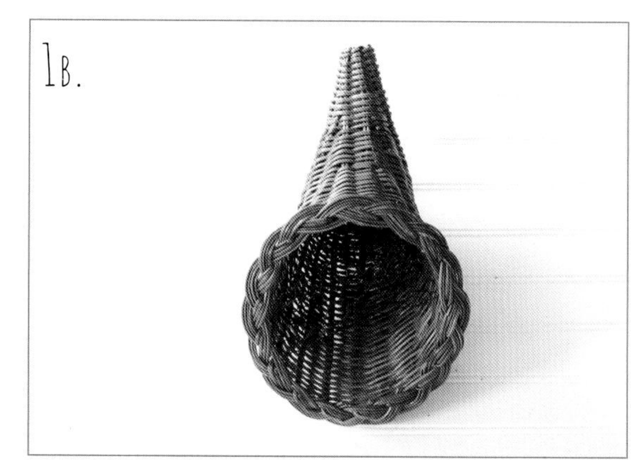

TWO

Attach the cornucopia to the grapevine wreath using 2 to 3 pieces of the 18-gauge floral wire stems. Insert the wire stems into the cornucopia around the grapevine wreath and twist the wire together on the back to secure. Attach the wire hanger to the top of the grapevine wreath (p. 5). Hang up and continue to work.

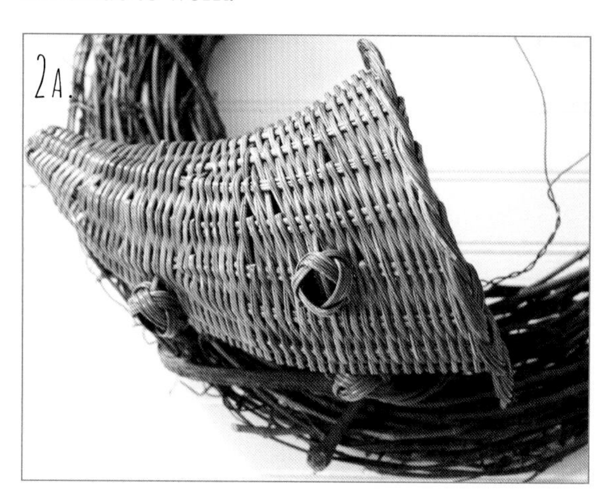

THREE

Clip the greenery and twig bushes apart, separating the greenery from the twigs and leaving a 2" stem. Trim the large pine cone and berry stems into 6–8" pieces, leaving a 2" stem.

FOUR

Apply hot glue to all of the twig stems and insert them into the grapevine form around the cornucopia.

FIVE

Apply hot glue to all but two of the greenery stems and insert them into the grapevine form. Arrange them on top of the twigs.

SIX

Insert two of the greenery stems into the cornucopia. Slightly bend the ends of the stems so they will catch the chicken wire.

SEVEN

Place two of the pine cone and berry stems into the cornucopia. Slightly bend at the end so they will catch to the chicken wire. Apply hot glue to the end of the third pine cone and berry stem and place it above the cornucopia.

EIGHT

Slightly bend the ends of the gourd picks and insert them into the chicken wire in the cornucopia.

NINE

Apply hot glue to the sedum picks and place them on the wreath to fill in any empty spaces.

TEN

Hang on your front door or in your kitchen—an all-season reminder of the bounty of fall.

Fall Leaf Wreath

I remember one fall afternoon, a few years back. The front yard was covered in leaves. It occurred to me that the girls were now old enough to help out with the chore of raking leaves—and that we might be able to have a bit of fun doing it. I ran down to the local hardware store and bought three new rakes. We spent the afternoon raking piles of leaves for jumping and eventually got those leaves to the curb. This wreath reminds me of that sweet afternoon.

Supplies

2 yellow leaf stems

2 red and orange leaf stems

1 green and purple leaf bush

Wire hanger (p. 5)

18" grapevine wreath

Tools

Wire snips

Hot glue gun

Hot glue

Instructions

ONE

Clip all of the leaf stems and bushes apart until each piece is between 4–6" long with a 1–2" stem at the end.

TWO

Attach the wire hanger to the top of the grapevine wreath (p. 5) and hang it up. Continue working. Begin with the yellow leaf stems. Apply hot glue to the end of the stems and insert them into the grapevine wreath, spacing them out evenly.

THREE

Apply hot glue to the ends of the green and purple leaf stems and place them between the yellow leaf stems.

FOUR

Apply hot glue to the ends of the red and orange leaf stems and place them between the yellow and green.

FIVE

Add more of the yellow leaf stems, spacing them out evenly. Then, add more of the red and orange leaf stems, spacing them out evenly. This layering of colors makes the wreath look more realistic.

SIX

Fill in any gaps with the rest of the leaf stems. Make sure to place the leaves in such a way that the wreath keeps its basic round shape.

Adjust the leaves so that you can see all the layers of colors.

SEVEN

Hang this wreath on your garden gate to welcome this beautiful season!

Eucalyptus Wreath

My family visits an orphanage in Mexico a couple of times a year. It is a dear place to us. On the property, there are towering eucalyptus trees that make me want to climb their branches and clip stem after stem and bring them home! But since this isn't possible, I visit my local craft store instead to create eucalyptus wreaths that remind me of that special place.

Supplies

3 packages dried eucalyptus

Wire hanger (p. 5)

24" wispy twig wreath

Tools

Wire snips

Hot glue gun

Hot glue

Instructions

ONE

Trim each of the dried eucalyptus stems to about 5–6" with a 1" stem on the end.

TWO

Attach the wire hanger to the wreath (p. 5). Hang up the wreath and continue working.

THREE

Apply hot glue to the ends of the eucalyptus stems and insert them into the sturdy inner circle of the wreath form.

FOUR

Continue to hot glue the eucalyptus stems, following the pattern of the twigs on the wreath form.

FIVE

When you have completed the circle, go back and add extra eucalyptus stems where there are gaps.

SIX

This wreath is simple enough to be enjoyed all year long. It works best indoors. Hang it on a mirror or work it into a gallery wall. You can also add bows for different seasons and occasions.

Trim Wreath

My youngest, Lila Mae, is an amazing self-taught weaver. She is forever walking up and down the yarn aisles of every craft store we set foot in. She has recently started adding new textures to her weavings, which has us branching out to the trim aisles, and I'm so glad we did! This whimsical, loopy trim and these giant pom-poms cried out to me, and I knew they needed to be my next wreath. And—it would be perfect on my weaver's bedroom door.

Supplies

8 yards of your favorite fringe trim

10" round Styrofoam wreath form, with rounded edges

Straight pins with ball heads

2" pom-pom trim (I bought enough to get 16 pom-poms)

1 yard matching ½" flat trim

Tools

Scissors

Instructions

ONE

Attach the end of the fringe trim to the Styrofoam wreath using the straight pins. Start the trim at the back of the wreath, and use several straight pins to secure it into the Styrofoam.

TWO

Wrap the fringe trim around the Styrofoam wreath, slightly overlapping the fringe to completely cover the Styrofoam.

THREE

Keep wrapping the trim around the Styrofoam until you cover the wreath. There is no need to add more straight pins while you wrap. Once you reach the end of the fringe trim, use several straight pins to secure the trim to the Styrofoam wreath. You can adjust the fringe trim at this point to make sure that all of the white Styrofoam is covered.

FOUR

Cut off each pom-pom, leaving it attached to the base trim with about a ½" section of base trim.

FIVE

Attach the pom-poms to the Styrofoam wreath by inserting the straight pins through the base of the trim and the Styrofoam.

SIX

Loop the ½" flat trim through the inner circle of the wreath and tie a knot at the top of the trim. Hang the wreath from the ½" flat trim. Hang up the wreath in an upright position and continue working.

Add more pom-poms and space them out evenly. Make sure to add pom-poms to the sides as well.

SEVEN

Give this wreath to your favorite weaver or seamstress! It would also be precious in a child's bedroom. You can get very creative with the colors and trims on this one.

Winter

"I wonder if the snow loves the trees and fields, that it kisses them so gently? And then it covers them up snug, you know, with a white quilt; and perhaps it says, 'Go to sleep, darlings, till the summer comes again.'"

—Lewis Carroll

I must admit that cold weather is not my favorite. But what that cold weather does for my family is my favorite. It makes us all come in a little earlier, gather around the fireplace, and watch movies together as we cuddle up on the big sectional. Our kitchen table stays covered in craft projects. We make and buy presents for our friends and family, wrap them up, and stack them under the tree. We welcome in the new year with friends. And while winter is the season that I want to come and go the quickest, there is something cozy about this time of hibernation. These wreaths add the spark needed during these colder, grayer days.

Winter Floral Wreath

I've worn the same earrings every day for years—giant white gold hoops. I actually feel like something is missing from my face without them on. I feel the same way about not having a wreath on my front door—it just feels like something is missing. So when the Christmas wreath comes down and it isn't quite time for the spring wreath yet, this wreath is the perfect answer!

Supplies

26-gauge floral paddle wire

18" grapevine wreath

Frosted eucalyptus garland

Wire hanger (p. 5)

3 white rose stems

1 lamb's ears stem

1 cream berry and greenery bush

1 cedar stem

1 eucalyptus bush

3 white berry stems

Wired linen pom-pom bow (p. 10)

Tools

Wire snips

Needle-nose pliers

Hot glue gun

Hot glue

Instructions

ONE

Twist the end of the 26-gauge floral paddle wire around a piece of grapevine just left of the top-center of the wreath to secure. Then, attach the frosted eucalyptus garland by wrapping the paddle wire around the garland and the wreath. Continue wrapping the paddle wire around the garland in a clockwise direction until the bottom section of the wreath is almost covered. Clip the excess garland here.

Secure the excess garland with the paddle wire about 2" from the original starting point just left of the top-center of the wreath. Working in a counterclockwise direction, wrap the paddle wire around the garland and wreath until you reach about 3" from the end of the first piece of garland. Clip the garland and secure the paddle wire to the back of the wreath by twisting the wire around a few of the grapevine pieces. When this is done, you'll see that you have left two gaps in the garland—in the upper-left corner and in the lower-left corner for the linen pom-pom bow and a cluster of two roses respectively.

1.

Clip the excess garland into individual pieces. Attach the wire hanger to the wreath (p. 5) and hang it up. Continue working.

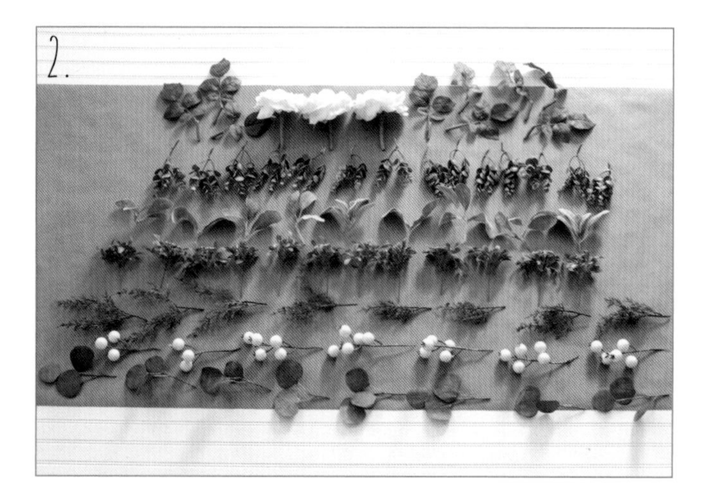
2.

TWO

Clip the white rose stems, lamb's ears stem, cream berry and greenery bush, cedar stem, eucalyptus stem, and white berry stems until they are about 3–4" long with a 1–2" stem on the end. Clip the leaves apart from the white rose stems.

THREE

Attach the linen pom-pom bow to the upper-left corner of the wreath in the gap. Use the wire from the bow and insert it into the grapevine and twist at the back to secure

FOUR

Apply hot glue to the ends of the white rose stems and insert into the grapevine. Cluster two of them together in the bottom-left corner of the wreath and the other in the top-right corner of the wreath.

FIVE

Apply hot glue to the ends of the leaves from the rose stems and insert them into the grapevine. Cluster them tightly around the white roses.

Apply hot glue to the extra frosted eucalyptus garland pieces and insert into the grapevine to fill in any gaps.

SIX

Apply hot glue to the ends of the lamb's ears pieces and insert them into the grapevine. Space them evenly.

SEVEN

Apply hot glue to the ends of the cream berry and greenery bush pieces and insert them into the grapevine. Space them evenly.

EIGHT

Apply hot glue to the ends of the cedar pieces and insert them into the grapevine. Space them evenly. Place some on the outer edge and inner edge of the grapevine wreath.

NINE

Apply hot glue to the ends of the white berry stems and insert them into the grapevine, spacing them evenly.

TEN

Apply hot glue to the ends of the eucalyptus bush pieces and insert them into the grapevine to fill in the last gaps.

ELEVEN

Hang this wreath on your front door during those "Christmas is over, but spring isn't here yet" months!

Snowball Wreath

One of my family's favorite movies is *Elf*. The snowball fight scene at the beginning of the movie that allows Elf to earn the respect of his brother and starts them off with a most magical afternoon is one of the best parts of the movie! Those brothers bonded over snowballs.

Supplies

6 (2½") Styrofoam balls

12" Styrofoam wreath with rounded edges

12 (2") Styrofoam balls

14 (1½") Styrofoam balls

20 (1") Styrofoam balls

Decoupage brush

Decoupage glue

Artificial snow

Red satin classic bow (p. 7)

1 yard matching red satin 2" ribbon

Tools

Hot glue gun

Hot glue

Scissors

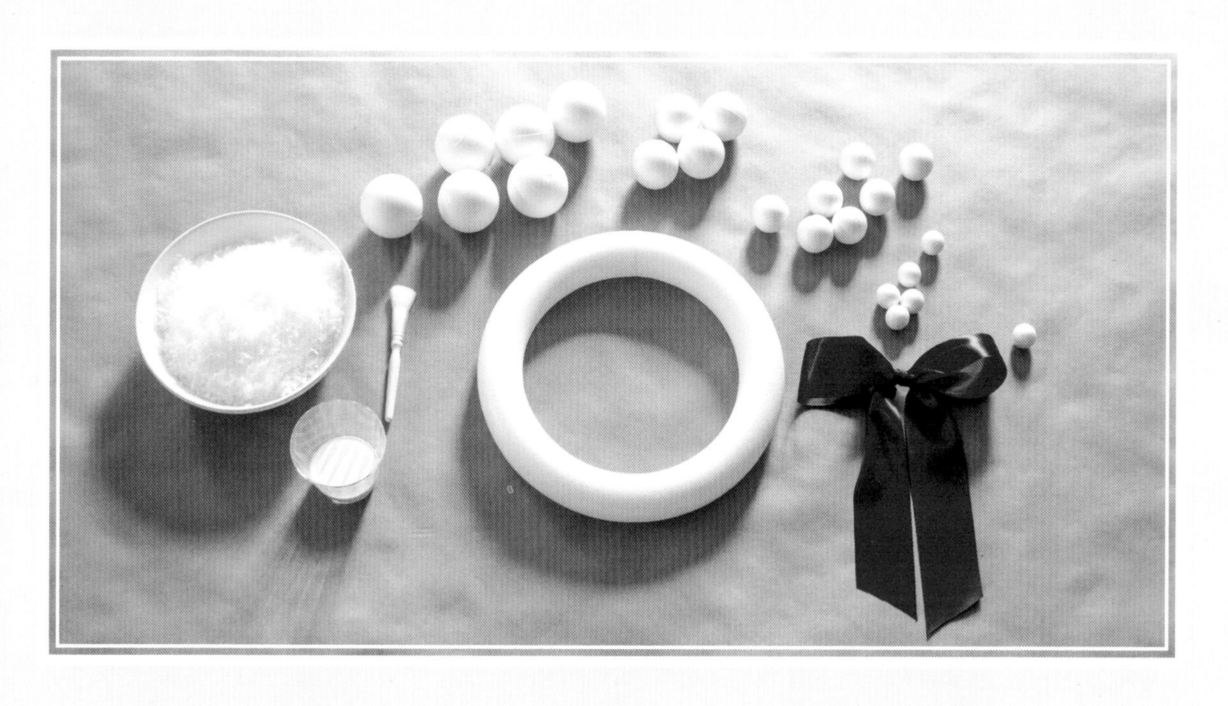

Instructions

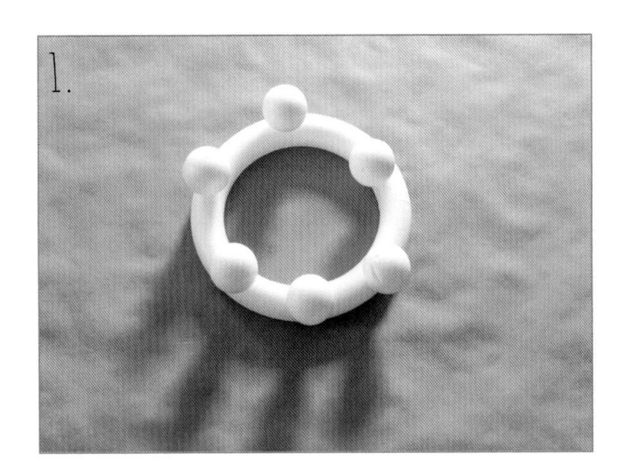

ONE

Apply hot glue to each of the 2½" Styrofoam balls and hold them in place on the Styrofoam wreath until the glue is dry. Space them evenly.

TWO

Apply hot glue to the bottom of the 2" Styrofoam balls and also the sides—you should be placing them so they are touching the larger Styrofoam balls. Hold them in place until the glue is dry.

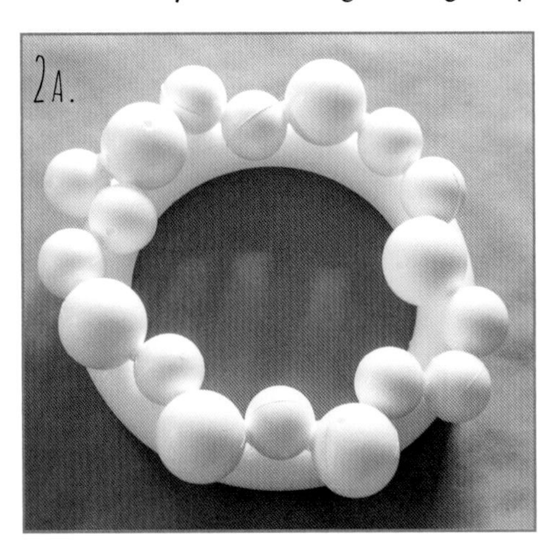

THREE

Apply hot glue to the bottom of the 1½" Styrofoam balls and also the sides—wherever they are touching the other balls—and hold them in place until the glue is dry. Space them evenly.

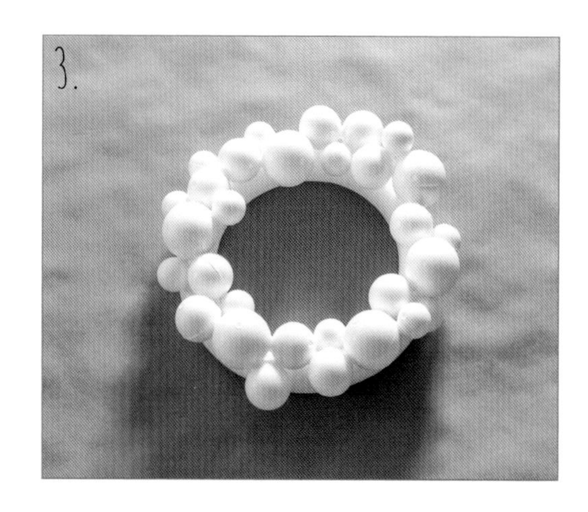

FOUR

Apply hot glue to the 1" Styrofoam balls and also the sides—wherever they are touching the other balls—and hold them in place until the glue is dry. Space them evenly. Use these to fill in gaps and to restore the round look of the wreath.

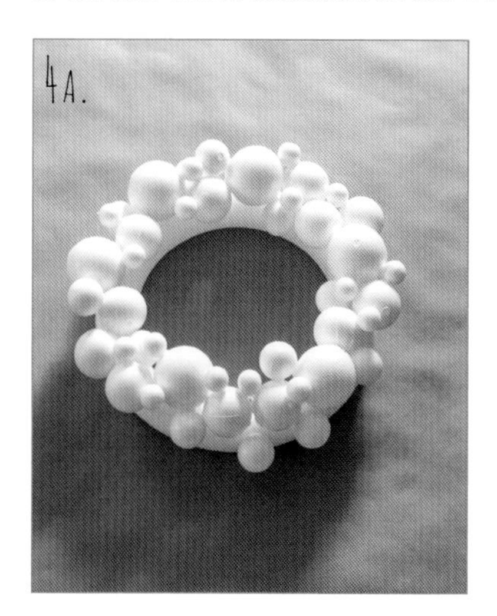

FIVE

Using the decoupage brush, apply a very thick, messy layer of decoupage glue to the tops of all of the Styrofoam balls.

SIX

Generously sprinkle the artificial snow over the entire wreath.

SEVEN

Allow it to dry for about 15 minutes, and then lightly shake the excess artificial snow off.

EIGHT

Apply hot glue to the back of the red satin classic bow and add it to the upper-right side of the wreath.

NINE

Loop the 2" red satin ribbon through the center of the wreath. Tie a knot at the top of the ribbon, and hang the wreath from the ribbon. Leave this one out all winter long!

Snowman Wreath

My girls dream of snowstorms, school closing, snowman building, and sledding down giant hills. Those dreams rarely come true where we live, but that does not stop them from flushing ice cubes down the toilet, wearing their pajamas inside-out, and sleeping with spoons under their pillows—all the superstitions they have heard that can encourage a snowstorm. Until then, we will have to settle for this precious fake snowman.

Supplies

Chunky white yarn

10" Styrofoam wreath with rounded edges

12" Styrofoam wreath with rounded edges

14" Styrofoam wreath with rounded edges

Craft foam top hat

2" red wired ribbon

Tools

Scissors

Hot glue gun

Hot glue

Instructions

ONE

Tie a knot in the chunky white yarn around the 10" Styrofoam wreath and begin wrapping the yarn around the Styrofoam wreath.

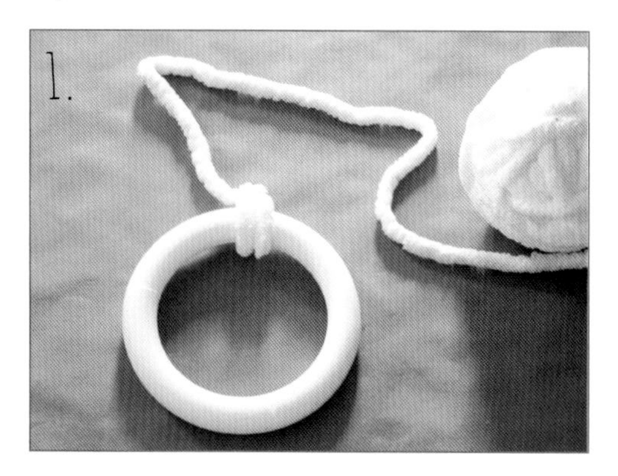

TWO

Tie a knot in the end of the yarn when you have finished covering the Styrofoam wreath. Tuck the end of the yarn under a piece of the wrapped yarn.

THREE

Wrap the 12" Styrofoam wreath and the 14" Styrofoam wreath with yarn in the same way.

FOUR

Cut 2 pieces of yarn about 8" long and use them to tie the wreaths together, with the 10" wreath at the top, the 12" wreath in the middle, and the 14" wreath on the bottom.

FIVE

Using scissors, cut the craft foam hat in half from the top.

SIX

Apply hot glue to the inside of the craft foam hat and glue it to the top of the 10" wreath.

SEVEN

Using the 2" red wired ribbon, tie a simple bow (like how you'd tie a shoelace) between the 10" and 12" wreath to act as the snowman's scarf.

EIGHT

Loop a piece of yarn to the top of the 10" wreath and tie a knot in the yarn. Using this yarn loop, hang the snowman wreath on your child's bedroom door and dream of the snow to come!

Cinnamon Stick and Dried Orange Slice Wreath

For Christmas, my grandmother always gave me a bag of oranges along with my presents. I always enjoyed the oranges, and finally I got curious enough to ask her why she always gave us oranges. She explained that when she was growing up, oranges were a huge treat, and she always received one in her stocking on Christmas morning. It was one of the things she looked forward to the most. Oranges at Christmas have been special to me ever since. This wreath is for my Muz and the bags and bags of oranges she shared with her grandchildren throughout the years.

Supplies

3 cedar stems

3 pine stems

3 spruce stems

Wire hanger (p. 5)

18" grapevine wreath

5 pine cones on sticks

24-gauge brown floral wire stems

26-gauge floral paddle wire

15 (5–6") cinnamon sticks

30 dried orange slices

½" red striped ribbon

Tools

Wire snips

Hot glue gun

Hot glue

Scissors

Instructions

ONE

Clip each of the cedar, pine, and spruce stems into 5–6" pieces with a 1–2" stem on the end.

TWO

Attach the wire hanger to the grapevine wreath (p. 5) and hang it up. Continue working. Apply hot glue to the ends of the cedar, pine, and spruce stems and begin adding them to the grapevine wreath.

THREE

Alternate the cedar, pine, and spruce stems as you go along the wreath. Continue adding the cedar, pine, and spruce stems until the grapevine wreath is completely covered.

FOUR

Apply hot glue to the pine cone sticks and insert them into the grapevine wreath. Space them out evenly. Turn the wreath over and trim the excess pine cone stick.

FIVE

To further secure the pine cones, wrap the 24-gauge brown floral wire stem around the pine cones and insert the wire into the grapevine wreath. Twist the wire to secure it.

SIX

Attach the 26-gauge floral paddle wire to a section of the grapevine wreath and wrap all of the cedar, pine, and spruce stems to the grapevine wreath to further secure it. Make sure not to wire any of the greenery down by only catching the stems with the wire.

SEVEN

Bundle three of the cinnamon sticks together and twist the 24-gauge brown floral wire stem around them to secure. Stack three slices of dried oranges together and run the brown floral wire stem through the tops of the oranges and twist the wire together to secure them.

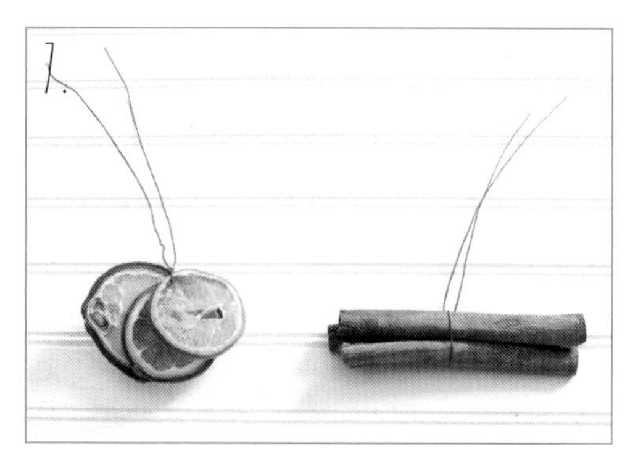

EIGHT

Using the ½" red striped ribbon, tie a knot around the cinnamon stick bundles.

NINE

Attach the cinnamon stick bundles to the wreath by inserting the 24-gauge brown floral wire stems into the grapevine wreath and twisting it together at the back. Space the cinnamon stick bundles evenly between the pine cones.

TEN

Attach the dried orange bundles to the grapevine wreath by inserting the wire into the grapevine wreath and twisting it together at the back. Space the orange bundles evenly between the cinnamon bundles and pine cones. Make sure to place the dried orange bundles at different angles.

ELEVEN

Hang this wreath in the kitchen to enjoy for the entire winter season.

Pine Cone Wreath

I love a project that is as quick as it is beautiful! This wreath can be made in less than 30 minutes. If you got out your stopwatch, you would probably amaze yourself at how fast this wreath comes together. Tie a simple bow on it and use for the entire fall and winter!

Supplies

40 triple pine cone clusters on stems

14" grapevine wreath

1½" cream wired ribbon

24-gauge brown floral wire stem

Wire hanger (p. 5)

Tools

Needle-nose pliers

Scissors

ONE

Insert the stem of the triple pine cone cluster into the grapevine wreath.

TWO

Use the needle-nose pliers to bend the end of the wire stem around a few of the grapevine pieces.

THREE

Add more of the pine cone clusters. Attach them very closely together. Make sure to cover the sides of the wreath as well.

FOUR

Tie a classic bow with the 1½" cream wired ribbon (p. 5). Attach it to the pine cone wreath by inserting a 24-gauge brown floral wire stem through the knot of the bow and inserting the wire into the wreath and twisting the wire at the back. Attach the wire hanger to the top of the wreath (p. 5).

FIVE

Make several of these as gifts for friends!

Winter Greenery and Frosted Pine Cone Wreath

In my Southern town, frost is a much more likely occurrence than snow. Layers of frost mound up on our rooftops and windshields many a winter morning. My chivalrous husband makes it his job to warm and scrape all of the cars that will be in use on any particular morning. There is never any snow shoveling at our house, but there is plenty of frost scraping. This frosty wreath looks far more at home on my front door than any of the snow-covered options!

Supplies

2 cedar stems

2 frosted greenery and pine cone stems

1 eucalyptus bush

6 frosted pine cones on sticks

26-gauge floral paddle wire

18" grapevine wreath

Wire hanger (p. 5)

Tools

Wire snips

Hot glue gun

Hot glue

Needle-nose pliers

Instructions

ONE

Clip each of the cedar stems, frosted greenery and pine cone stems, and eucalyptus bush into 6–10" pieces with 1–2" stems at the end. Remove the pine cones from the frosted greenery stems. Clip the sticks on the frosted pine cones to about 6".

TWO

Attach the 26-gauge floral paddle wire to the grapevine wreath by twisting the wire to a piece of grapevine.

THREE

Attach the wire hanger to the grapevine wreath (p. 5). Hang it up, and continue working so you can make sure that the greenery hangs correctly. Attach the first 4 pieces of frosted greenery and pine cones to the lower half of the grapevine wreath by inserting the stems into the grapevine and wrapping the floral paddle wire around the stems and the grapevine wreath.

FOUR

Continue adding the frosted greenery and pine cone stems until the bottom is full.

FIVE

Insert the ends of the cedar stems into the grapevine wreath and wrap the floral paddle wire around the stems and the grapevine wreath. Bend some of the pieces to follow the shape of the grapevine wreath, but let others stick out to the side.

SIX

Attach the eucalyptus stems on top of the other greenery by inserting the ends into the grapevine wreath and wrapping the floral paddle wire around the stems and grapevine wreath. Clip the paddle wire and secure it at the back by twisting it to a piece of the grapevine.

SEVEN

Apply hot glue to the ends of the frosted pine cone sticks and insert them into the grapevine wreath.

EIGHT

Apply hot glue to a few more of the eucalyptus stems and insert them into the grapevine wreath to fill out the area around the frosted pine cones.

NINE

Hang this on the front door to welcome your friends and family all winter long!

Gumdrop Wreath

My Anna Belle was sixteen months old and just beginning to toddle around when her second Christmas rolled around. We had a Christmas tree in our kitchen, which had twigs with real gumdrops on them sticking out randomly from the tree. I began to notice sticky spots on my white kitchen floor but didn't think much of it; sticky floors were my norm. Then I began to notice that some of the gumdrops were no longer sugarcoated, but looked shiny instead. One afternoon, I came into the kitchen and there stood my little Anna Belle, sucking all of the sugar off one of those gumdrops, while her drool ran down her little chin and onto my floor. Mystery solved. I have loved gumdrops ever since.

Supplies

290 wooden toothpicks

12" Styrofoam wreath with rounded edges

140 small gumdrops

120 large gumdrops

2½" red gingham ribbon

Tools

Wire snips

Needle-nose pliers

Scissors

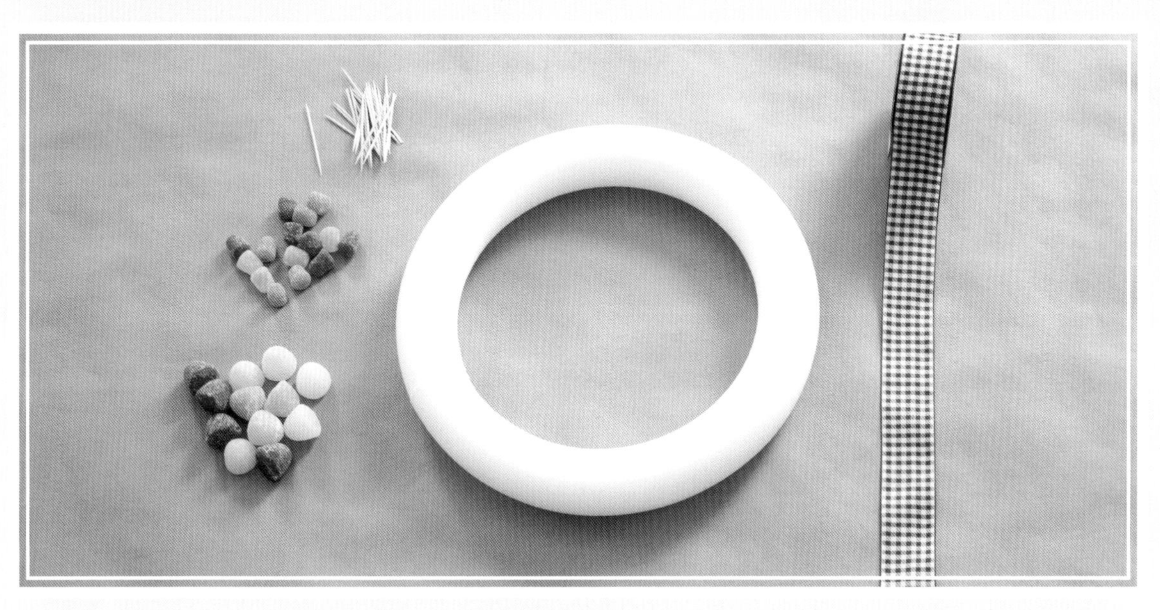

Instructions

ONE

Clip the toothpicks into thirds for the small gumdrops and in half for the large gumdrops. Using needle-nose pliers, insert the ends of the toothpicks into the center of the flat bottoms of the gumdrops.

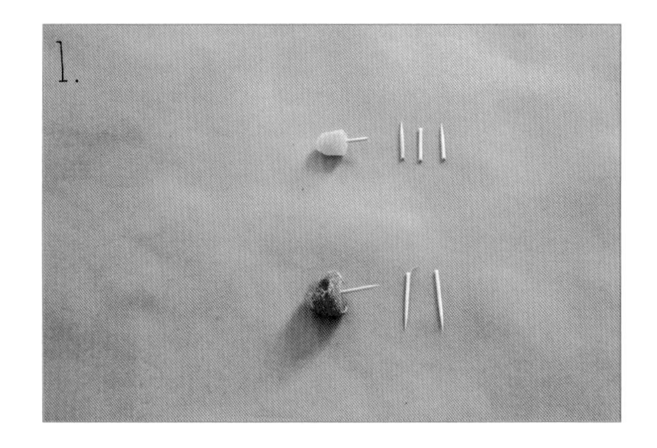
1.

TWO

Insert the gumdrop toothpicks into the inner circle of the Styrofoam wreath. Alternate between the large gumdrops and small gumdrops.

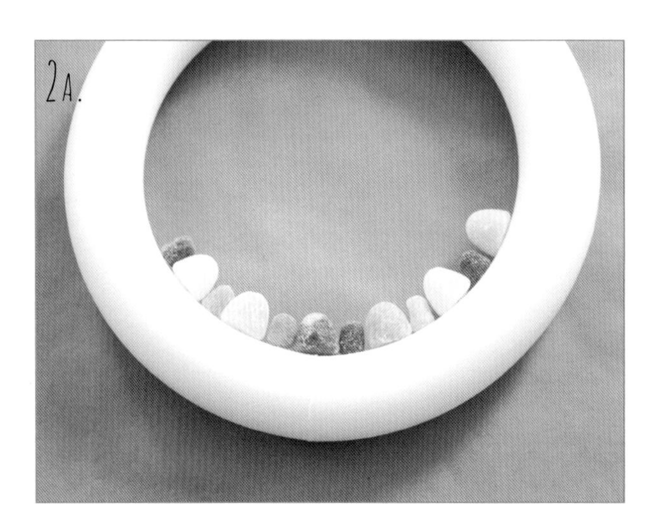
2A.

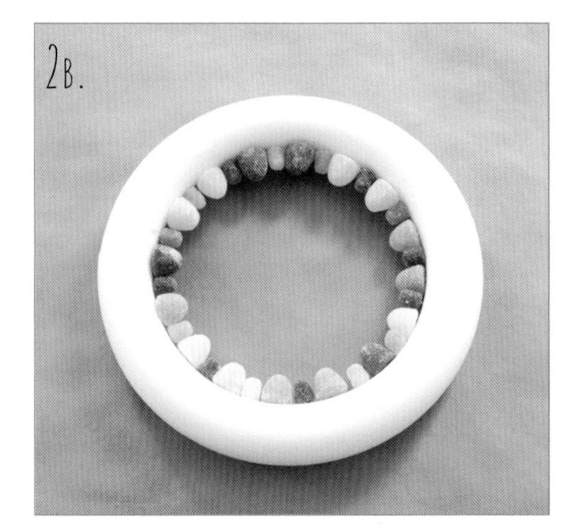
2B.

THREE

Begin randomly inserting the large and small gumdrops around the wreath. Space out the colors evenly. Continue adding gumdrops until the wreath is covered.

FOUR

Loop the 2½" red gingham ribbon around the top of the wreath and tie a simple bow (like how you would tie a shoelace) at the top.

Hang this wreath high enough so that your toddlers (and pets) can't partake of the sugar! If you notice that any of the gumdrops are shiny and no longer sugarcoated, you will know what happened!

Christmas Tree Wreath

We have always been an artificial tree family because it's just more practical. One day, though, I'd like to have the entire Christmas tree experience. Walk through the woods, pick out the perfect tree, chop it down, tie it to the top of the car, and take it home. This wreath reminds me of a walk through a Christmas tree forest.

Supplies

1½" green satin ribbon

Floral pins

14" straw wreath

20-gauge green floral wire stems

35–40 bottle brush trees, in varying sizes and colors

Tools

Wire snips

Needle-nose pliers

Scissors

Instructions

ONE

Pierce the end of the 1½" green satin ribbon with the floral pin and insert it into the back of the straw wreath.

1.

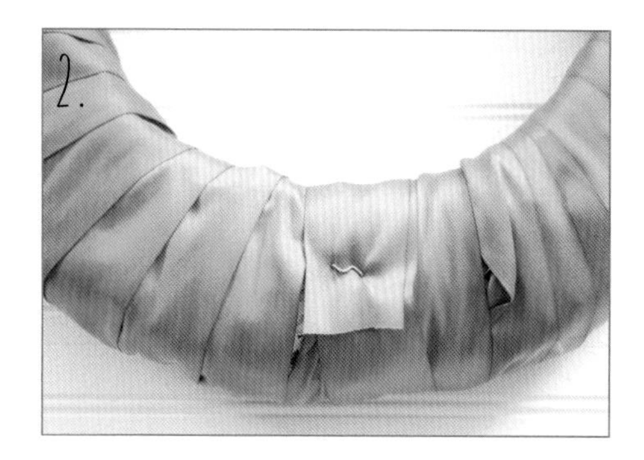
2.

TWO

Wrap the entire 14" straw wreath with the green satin ribbon, overlapping the ribbon so the straw wreath is completely covered. Use another floral pin to secure the ribbon at the end.

THREE

Use the 20-gauge green floral wire stems to attach the largest bottle brush trees to the two bottom corners of the wreath. Secure them by twisting the wire together at the back.

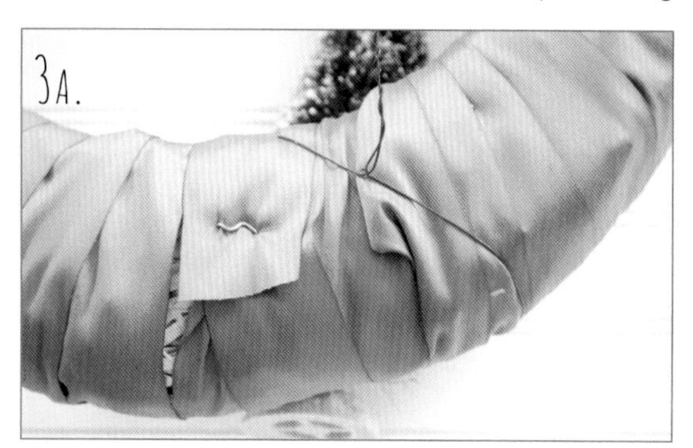
3A.

3B.

FOUR

Loop a piece of the green satin ribbon through the center of the wreath and tie a knot in the ribbon. Hang the wreath by the ribbon loop and continue working.

Begin by attaching the largest trees and then work your way down to the smallest trees as you go. Hot glue the trees that are the next size down beside the largest trees, and so on and so forth, until you have a variety of sizes and colors. When you firmly press the glued trees to the wreath, you will find that the bristles of the trees will allow them to mesh together.

FIVE

Hang this tree wreath and enjoy the Christmas season!

Nontraditional Wreaths

"You can't use up creativity. The more you use, the more you have."
—Maya Angelou

After you make a few wreaths, you'll begin to see wreath potential in just about everything. Just ask yourself a few questions: Is it round? Can it hold flowers? Can I attach flowers to it? Is it about the size of a wreath? Can it be hung? Is it interesting to look at? Get creative—the possibilities are endless!

Vintage Window Screen Hanging

There is something about a screened porch that makes you breathe a little deeper and slow down for a bit. Whether you are sipping sweet tea, enjoying a cup of coffee, soaking in some solitude, or enjoying the company of friends and family, the screened porch invites a certain peace. This special place is the inspiration for this little hanging.

Supplies

Floral foam

Vintage wooden window screen

24-gauge brown floral wire stems

Floral pins

Preserved sheet moss

1 cotton stem

1 lamb's ears stem

Tools

Serrated knife

Wire snips

Needle-nose pliers

Scissors

Instructions

ONE

Using the serrated knife, cut a piece of floral foam to 3" x 5". Attach the floral foam onto the inside corner of the vintage wooden window screen using the 24-gauge brown floral wire stems. Twist the wire together at the back of the screen to secure.

TWO

Use the floral pins to secure the preserved sheet moss over the floral foam.

THREE

Clip the cotton stem and lamb's ears stem into individual pieces of varying sizes.

FOUR

Make a tiny loop with the brown floral wire at the top of the screen to hang the screen from. Hang it up and continue working.

FIVE

Insert the longer of the cotton stems and lamb's ears to the top of the floral foam.

SIX

Continue to fill up the floral foam with the cotton and lamb's ears.

SEVEN

Hang this in a place where you need to be reminded of quiet times on a screened porch.

Welcome

Hanging Chalkboard

My girls always enjoyed watching *Little House on the Prairie* when they were little. They loved Laura and Mary, and they were fascinated with the little schoolhouse and the slates they worked on and the bonnets they wore. So fascinated that we bought them their own little slates and bonnets. They would spend hours pretending to be in that schoolhouse. This hanging chalkboard is a tribute to those precious Little House days!

Supplies

Slate chalkboard with wood frame

Chalk

2" ribbon

1 lavender stem

½" coordinating ribbon

Tools

Scissors

Wire snips

Staple gun

3/8" staples

Instructions

ONE

Prime your chalkboard. Taking the side of a piece of chalk, color the entire chalkboard. Then, using a soft cloth, wipe all of the chalk off. This will prevent you from staining the chalkboard with very first thing you write on it.

TWO

Cut your 2" ribbon to ¾ of a yard. Fold the ends of the ribbon to hide the raw edges.

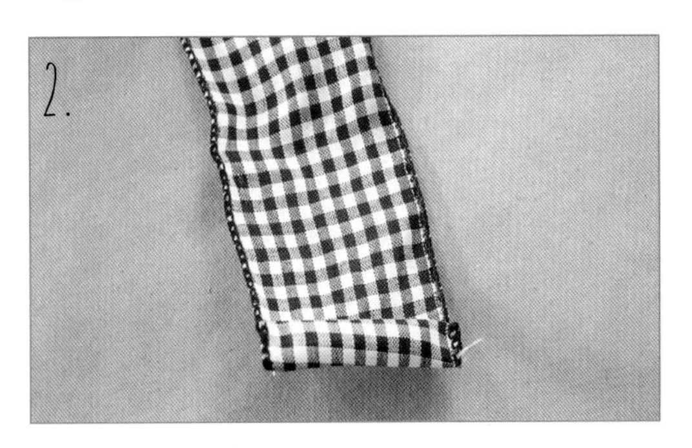

THREE

Fold down the long finished side of the 2" ribbon in half so they meet in the middle lengthwise. Then, fold the ends of the short edge twice more and staple the ends a couple of times to the top right edge of the wood frame to secure. Do the same thing to the other end of the ribbon and staple it to the top left edge of the wood frame. You will hang the chalkboard from this ribbon.

FOUR

Clip the lavender into several pieces of varying length. Leave a 1–2" stem on the ends. Make sure that none of the lavender pieces are taller than the height of the chalkboard.

FIVE

Gather the stems of lavender together to create a little bouquet and secure with the ½" ribbon. Tie a knot and follow with a simple bow (like how you would tie a shoelace).

SIX

Staple a few stems of the bouquet of lavender to the wooden side of the chalkboard. You will need to use 3 to 4 staples to make it secure.

SEVEN

Hang your chalkboard using the ribbon hanger and make sure the lavender bouquet is secure.

EIGHT

Your chalkboard is now ready to welcome guests or make announcements!

Ice Skate Arrangement

The first time I saw an ice-skating rink, I was 8 years old. It was planted right in the center of our new mall. I remember watching in awe as people in white skates glided gracefully by. I dreamed of owning a pair of skates and twirling on the ice and, a few weeks later, I had my first try! I realized rather quickly that there would never be any graceful gliding for me—but it didn't keep me from admiring those beautiful white skates.

Supplies

1 pine and pine cone bush

Vintage ice skates

2 cedar stems

Wired linen pom-pom bow (p. 10)

Tools

Wire snips

Instructions

ONE

Separate the pine and pine cone bush into several pieces, leaving the stems as long as you can.

1.

TWO

2.

Hang the ice skates up by their laces. Adjust the laces until the skates are staggered and stacked on top of each other. This will help them to lie flat against the door.

THREE

Bend the first cedar stem and insert it into the top skate. Bend the second cedar stem and insert it in the second skate.

3.

FOUR

Begin adding the pine stems to the top skate. Arrange each pine stem with the pine cone at the opening of the bottom skate. Cluster the pine cones closely together as shown.

FIVE

Use the wire to attach the linen pom-pom bow to a piece of greenery at the base of the skate. Keep the bow close to the pine cone cluster.

SIX

Fill in the skates with the rest of the pine stems. Hang this on a vintage sled or on a hook in your entry and enjoy it all winter!

Vintage Rake Hanging

When you have three cute teenage girls living in your house, you will occasionally have boys around, too. And one of these boys happened to be handy, even carrying around tools in the back of his truck. With one simple cut of his circular saw, this vintage rake was ready to become an adorable wall hanging.

Supplies

Artificial bittersweet stem

26-gauge floral paddle wire

Vintage rake, with the handle cut down to about 4"

Raffia

Raffia classic bow (p. 7)

Tools

Wire snips

Scissors

Hot glue gun

Hot glue

Instructions

ONE

Bend the end of the artificial bittersweet stem to create a hook.

1A.

1B.

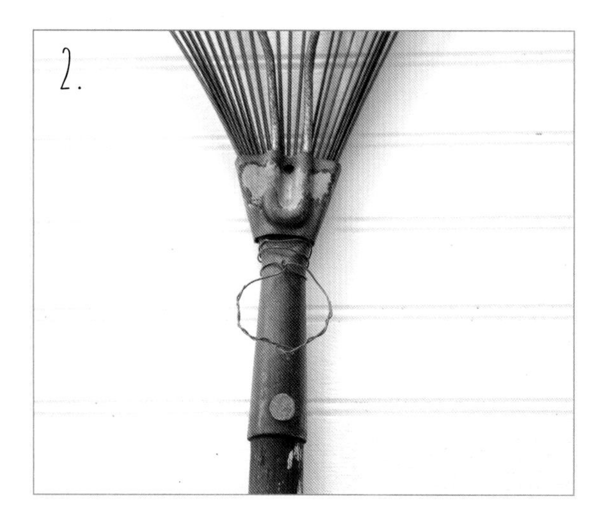
2.

TWO

Wrap the 26-gauge floral paddle wire around the rake several times where the metal meets the wood. Create a loop with the wire on the back of the rake. You will use this loop to hang the rake.

THREE

Twist the floral paddle wire around the hook you created on the end of the bittersweet stem.

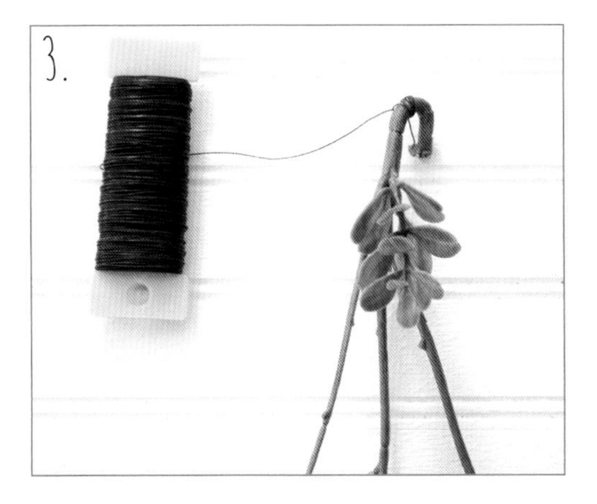

FOUR

Using the floral paddle wire, secure the bittersweet to the handle of the rake where the wood and metal meet. Tie a knot of raffia over the hook in the bittersweet.

FIVE

Let the raffia flow freely from the knot.

SIX

Hot glue the raffia classic bow over the wire and raffia knot. Hang this wreath in a place where you can welcome all of your male visitors if you, too, have teen girls living in your house!

Burlap Bag Hanging

Burlap might be the most versatile textile around. It is totally at home in the garden center, wrapped around the root ball of a tree; but it can also show up at a wedding as a table runner, and still look beautiful. This burlap bag is so easy to make and can be dressed up or down for any occasion. And there is no sewing required!

Supplies

2 yards 12"-wide natural burlap (sold by the roll with finished edges)

Clear craft glue

6–8 clothespins

Fray check

½" ribbon

2 safety pins

1 hydrangea and mixed greenery bush

Floral foam

Tools

Scissors

Wire snips

Instructions

ONE

Cut the burlap into two 1-yard pieces. Fold the first piece of burlap in half with the finished edges matching along the sides.

1.

TWO

Run a bead of clear craft glue down the finished edges of the sides of the burlap and glue together. Leave the top edge unglued to create the bag's opening.

2.

THREE

Use clothespins to secure the edges of the burlap together until the glue is secure. But don't wait too long—if the glue is allowed to dry totally before you remove the clothespins, they will be glued to the burlap and will be very hard to remove.

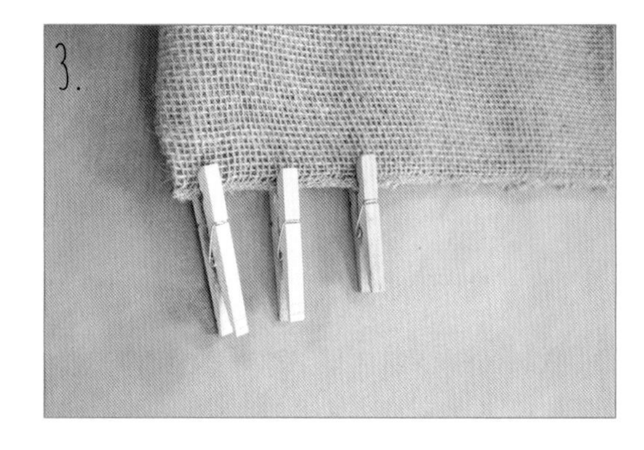
3.

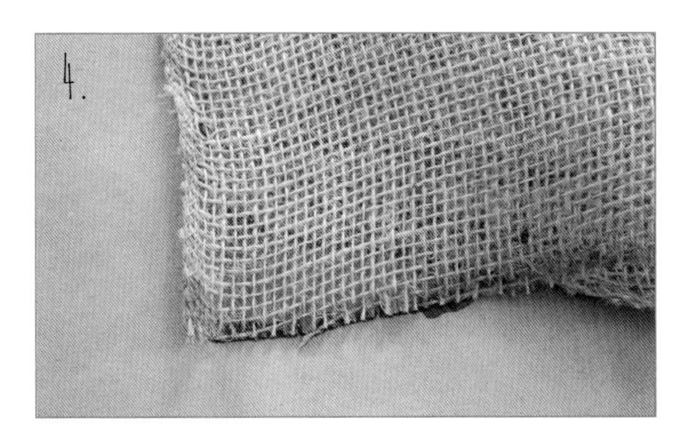

FOUR

Run a generous bead of fray check along the circumference of the top raw edge you have from cutting the burlap. This will keep the burlap from unraveling. Wait for the glue and fray check to dry completely before moving to the next step.

FIVE

You are going to create a second layer of burlap over the first piece to keep you from being able to see through the bag. Take the second piece of burlap and fold it in half over the first piece that you glued together, making sure to line up the edges. Apply a bead of clear craft glue along the finished side edges of the second piece and glue it to the finished side edges of the first. Secure with clothespins until the glue is secure, but don't wait too long, again.

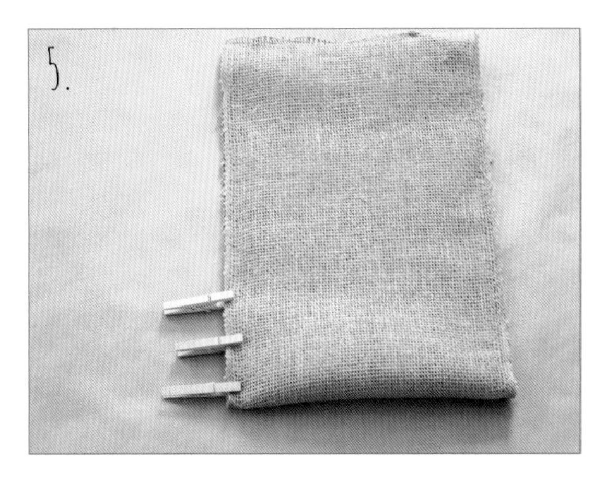

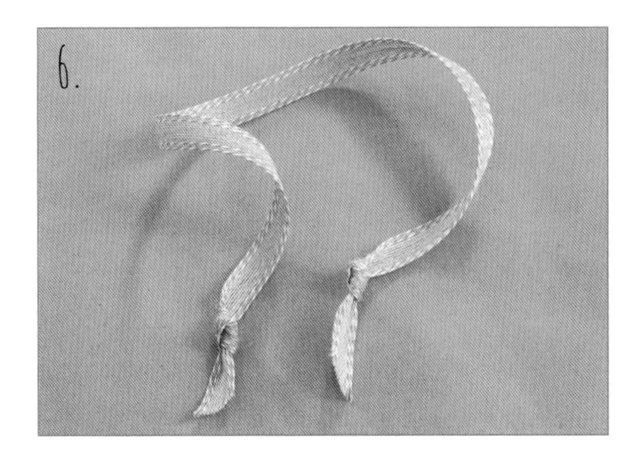

SIX

Cut a piece of the ½" ribbon about 1 yard long and knot each end of the ribbon.

SEVEN

Use the safety pins to attach the knots of the ribbon to the top edges of the burlap bag.

EIGHT

Clip the hydrangea bush into individual stems, with each piece about 8–10" long.

NINE

Fill the burlap bag with floral foam. Hang it up by the ribbon and continue working.

TEN

Fill the top of the bag with hydrangea stems so that the hydrangea blooms are just peeking over the top of the bag. Continue to add hydrangea blooms and stack them on top of one another.

ELEVEN

Add the greenery stems to the top and leave a piece trailing down the side of the burlap bag.

TWELVE

Add the last of the greenery pieces to fill in gaps. Dress this little beauty up or down as needed!

Canning Lid Wreath

There were a couple of summers when I had the vegetable garden of my dreams. A sweet older man at my church let me garden with him, and he taught me everything he knew from his years of tilling, planting, feeding, watering, and gathering. I spent hours working on a plot of land, and then my mom and I spent hours canning all of the bounty. This wreath reminds me of those special times.

Supplies

6" wooden embroidery hoop with screw

53 canning rings

3 yards 1½" ribbon

Tools

Scissors

Instructions

ONE

Remove the screw from the embroidery hoop. You will only be working with the outer circle of the embroidery hoop. Save the inner circle for another project!

1.

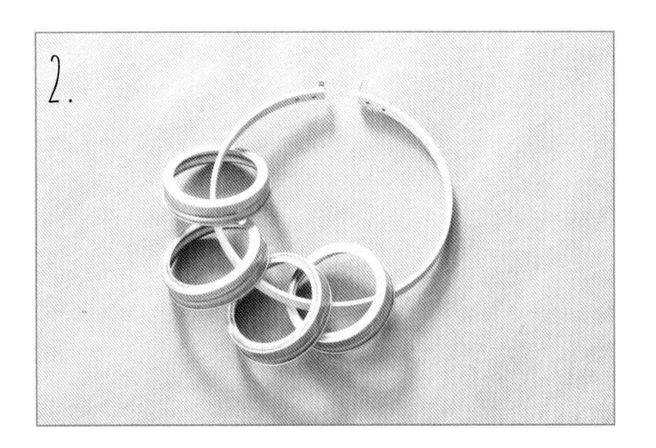
2.

TWO

Add the canning lids to the embroidery hoop. Face all of the canning lids in the same direction.

THREE

Pack the canning lids on as tight as you can. Screw the ends of the embroidery hoop back together.

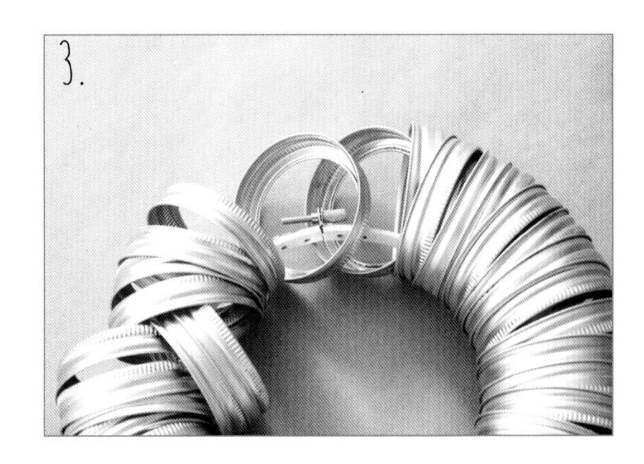
3.

FOUR

Loop the 1½" ribbon through the center of the wreath and tie a simple bow (like you would tie a shoelace) at the top. Use this ribbon loop to hang the wreath. Adjust the lids until they are laying down correctly. If you can't get them to stay put, add more lids.

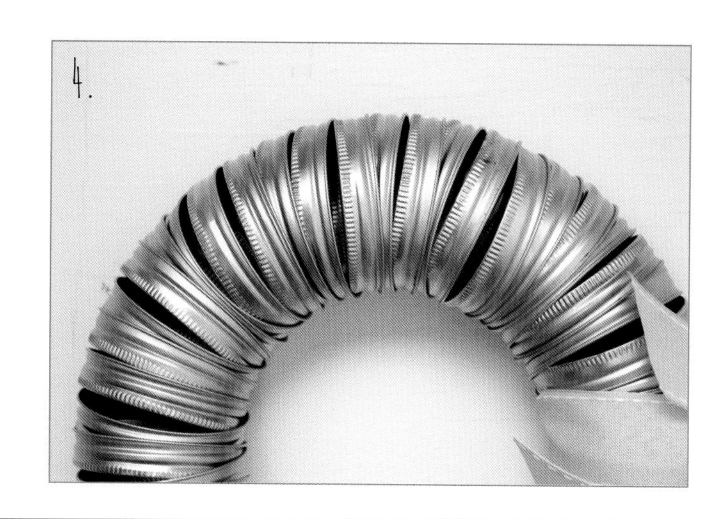

FIVE

Gift this to a friend who loves to garden and can, or enjoy on your pantry door!

Vintage Picture Frame

I began frequenting antique malls and junk stores several years back. I am always drawn to these beautiful old frames and can't resist bringing them home. Problem is, they never seem to fit any pictures I have! This is the perfect project for such a frame that you would love to display.

Supplies

2 cream ranunculus stems

1 large cream hydrangea stem

3 eucalyptus stems

Chicken wire

Vintage picture frame

26-gauge floral paddle wire

Tools

Wire snips

Scissors

Instructions

ONE

Clip each cream ranunculus stem, cream hydrangea stem, and eucalyptus stem so that the leaves are separate from the blooms and the stems are each about 4" long with a 2" stem. Some of the leaves will remain with the flowers.

1.

TWO

Using the wire snips, clip a piece of chicken wire 6" x 8" and roll into a shape that will fit the corner of the frame.

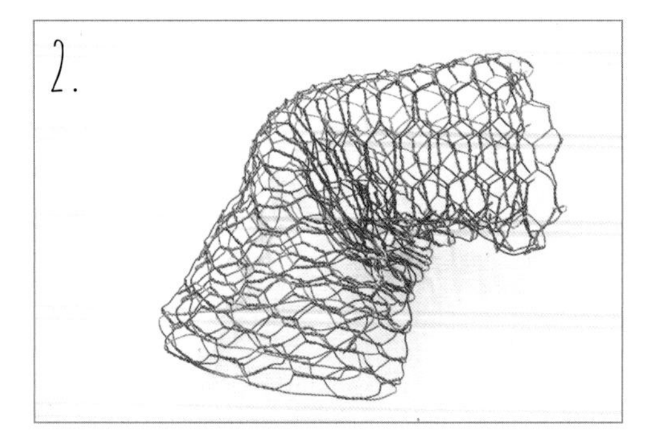
2.

THREE

Lightly press the chicken wire to the frame to get it to form around the corner. Use the 26-gauge floral paddle wire and wrap the chicken wire around the frame, securing it to the corner of the frame. When it is secure, clip the floral paddle wire and tie a knot in the end to make it stay.

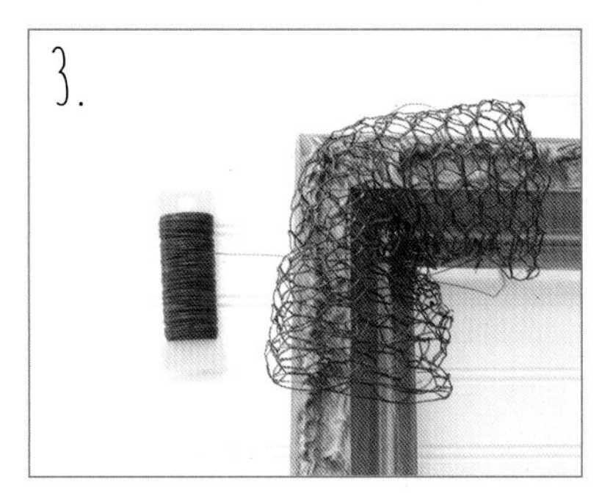
3.

FOUR

Add the large cream hydrangea bloom and its leaves to the middle of the corner of the frame. Slightly bend the end of the hydrangea stem so that it will stay secure in the chicken wire.

4.

FIVE

Add the leaves of the ranunculus around the hydrangea bloom. Keep the leaves balanced evenly on each side of the hydrangea bloom.

SIX

Begin adding the cream ranunculus stems to each side of the hydrangea bloom by inserting the stems into the chicken wire.

SEVEN

Fill in the gaps with the eucalyptus stems.

EIGHT

Next time you see a beautiful old frame, don't worry about being unable to fit your pictures into the frame. Buy it for its beauty and transform it into a wreath!

Vintage Dustpan Arrangement

My oldest, Daisy, is always up for a quick antique store run in the next town over. On one of our little excursions, I spotted this vintage dustpan hanging on the wall. I knew it would look adorable filled with succulents and hanging on a mudroom door. Then I thought of what my grandmother would think—she actually used one of these to clean her kitchen floors, and here I am decorating with it! She would have laughed and thought I was crazy! This one is dedicated to my Mama Lil.

Supplies

22-gauge floral wire stem

1 vintage dustpan

3 yards of 2" wired ribbon

Floral foam

4 succulents of varying sizes

1 fern bush

Tools

Serrated knife

Scissors

Wire snips

Instructions

ONE

To make a custom wire hanger for the dustpan, twist the 22-gauge floral wire stem around the handle of the dustpan (near the top), then twist the wire into a loop about 6" down the dustpan handle.

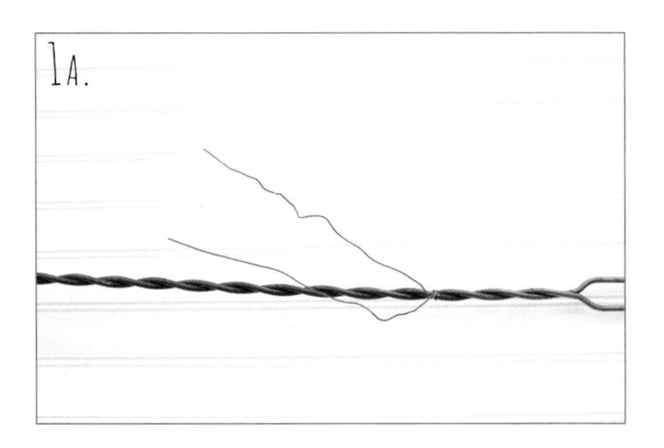

TWO

Hang the dustpan from the wire hanger. Using the 2" wired ribbon, tie a simple bow (like you would tie a shoelace) around the handle of the dustpan to hide the wire hanger.

THREE

Fit the floral foam into the opening of the dustpan. My block of foam happened to fit perfectly, but you can easily cut the foam with a serrated knife if needed to fit.

FOUR

Start with the largest succulent and insert it in the foam just left of the center.

FIVE

Secure the next succulent beside the first on the left.

SIX

Add the third succulent to the far right.

SEVEN

Add the fourth succulent between the first and third succulents you added.

EIGHT

Take the fern bush apart so you can work with each individual fern frond.

NINE

Add the fern by inserting the stems into the foam. Add them on top of the succulents and trail a couple off the side on the far left.

TEN

Find the perfect home for this wreath in a laundry room or mudroom—and wait for you grandmother to laugh at you!

About the Author

Melissa Skidmore has made her home in Murfreesboro, Tennessee, for the last nineteen years and counting. She loves to be with her family more than anything else in the world. She is married to the love of her life, David Skidmore. Together, they have three beautiful girls—Daisy Sue, Anna Belle, and Lila Mae. She previously spent several years decorating the homes of friends and dabbling in the world of blogging, but she has recently turned her creative energy to wreath making. When she isn't making wreaths for her Etsy shop, DaisyMaeBelle, her favorite activity is to spend time on projects that will make her home more welcoming.